# Computational Intelligence

# RIVER PUBLISHERS SERIES IN MATHEMATICAL, STATISTICAL AND COMPUTATIONAL MODELLING FOR ENGINEERING

*Series Editor*

**MANGEY RAM**
*Graphic Era University,*
*India*

Applied mathematical techniques along with statistical and computational data analysis has become vital skills across the physical sciences. The purpose of this book series is to present novel applications of numerical and computational modelling and data analysis across the applied sciences. We encourage applied mathematicians, statisticians, data scientists and computing engineers working in a comprehensive range of research fields to showcase different techniques and skills, such as differential equations, finite element method, algorithms, discrete mathematics, numerical simulation, machine learning, probability and statistics, fuzzy theory, etc.

Books published in the series include professional research monographs, edited volumes, conference proceedings, handbooks and textbooks, which provide new insights for researchers, specialists in industry, and graduate students.

Topics included in this series are as follows:-

- Discrete mathematics and computation
- Fault diagnosis and fault tolerance
- Finite element method (FEM) modeling/simulation
- Fuzzy and possibility theory
- Fuzzy logic and neuro-fuzzy systems for relevant engineering applications
- Game Theory
- Mathematical concepts and applications
- Modelling in engineering applications
- Numerical simulations
- Optimization and algorithms
- Queueing systems
- Resilience
- Stochastic modelling and statistical inference
- Stochastic Processes
- Structural Mechanics
- Theoretical and applied mechanics

For a list of other books in this series, visit www.riverpublishers.com

# Computational Intelligence

**Editors**

**Sushil Chandra Dimri**
Graphic Era (Deemed to be) University,
Dehradun, India

**Lata Nautiyal**
University of Roehampton, London

**Akshay Kumar**
Graphic Era Hill University,
India

**Richa Indu**
Himalayan Institute of Technology Asthal Campus,
Dehradun, India.

River Publishers

Routledge
Taylor & Francis Group
NEW YORK AND LONDON

**Published 2026 by River Publishers**
River Publishers
Broagervej 10, 9260 Gistrup, Denmark
www.riverpublishers.com

**Distributed exclusively by Routledge**
605 Third Avenue, New York, NY 10017, USA
4 Park Square, Milton Park, Abingdon, Oxon OX14 4RN

*Computational Intelligence / Sushil Chandra Dimri, Lata Nautiyal, Akshay Kumar and Richa Indu.*



Routledge is an imprint of the Taylor & Francis Group, an informa business

ISBN 97887-438-1171-8 (hardback)
ISBN 97887-438-1276-0 (paperback)
ISBN 97887-438-1274-6 (Ebook)
ISBN 97887-438-1275-3 (E-pub)
ISBN 97887-438-1273-9 (Ebook master)
DOI: 10.1201/9788743812739

# Contents

# Preface

Computational intelligence is not limited to optimize complexities of complex computation but it has wider scope which includes development of efficient full proof generic mechanism to address the problems arises in field of science and technology.

In Computer science computational intelligence widely affects all aspects of designing of algorithms, system mechanism and schemes used in optimization. Computational intelligence is nature induced methodologies scheme or approaches which are used when basic mathematical reasoning leads to uncertainty and complex situations. Computational intelligence is little different from Artificial intelligence, Artificial intelligence is basically based on logic and rules whereas Computational intelligence based on nature inspired schemes.

The basic three component of Computational intelligence are Neural Networks, Fuzzy Systems and Evolutionary computation.

Neural network is inspired by human brain neuro system, which is a large parallel distributed systems with ability of adapt and learn. Fuzzy system is based on human language used to handle uncertainty and error full data. While Evolutionary computation is based on biological evolution like reproduction, mutation etc.

But now Computational intelligence is not limited to these three areas, machine learning methods, swarm intelligence, SVM, Chaotic system has been added to Computational intelligence. Decision making with Computational intelligence gives us more accurate precise outcome in less time and less complex.

This book discusses few Computational intelligence approaches used in various areas of computer science, and exhibits the power of Computational intelligence in problem handling and providing the solutions. In coming time, the methods of nature will provide the lights to the human being to deal with their problems in more efficient and simple manner.

Chapter 1: explores the use of Artificial intelligence and Computational intelligence in healthcare sectors, how diagnosis and treatment can be improved with the use of Computational intelligence.

Chapter 2: explores the use of Artificial intelligence and Computational intelligence in medical diagnostics. The use of artificial intelligence and machine learning have revolutionary potential, especially for atrial fibrillation or other heart-related disease.

Chapter 3: This chapter presents a machine learning-based approach for accurately predicting the resonant frequencies of dual-band circular patch antennas.

Chapter 4: The increasing complexity and diversity of cyber threats require the development of new detection and mitigation methodologies. This chapter discusses the integration of computational intelligence techniques, such as Quantum Neural Networks (QNN), Generative Adversarial Networks (GAN), Artificial Neural Networks (ANN), and other machine learning approaches to enhance cybersecurity systems.

Chapter 5: discusses use of intelligence in Agile Software Development.

Chapter 6: This chapter discusses the intelligent algorithms to deal with financial security and safe transactions.

Chapter 7: this chapter explores Artificial intelligence / Computational intelligence -enabled psychological treatment refers to innovative approaches that utilize artificial intelligence to provide psychological assistance. This model leverages natural language processing (NLP), machine learning (ML), and sentiment analysis to simulate human-like therapy conversations, delivering personalized and immediate psychological support.

Chapter 8: The chapter explores the convergence of artificial intelligence (AI), machine learning (ML), and block chain technology, outlining the evolutionary trajectory from conventional block chain systems to intelligent block chain architectures.

Chapter 9: Discusses use of intelligence in simulation of graphic entities on computer screen.

Chapter 10: This chapter will discuss various computational techniques for intelligence including artificial neural networks, quantum neural networks, and generative adversarial networks in strengthening cyber security frameworks.

Chapter 11: This chapter discusses cyber security risks pose substantial problems to preserving data integrity, privacy, and security in the continuously evolving context of financial transactions.

# List of Contributors

**Chauhan, Piyush,** *Symbiosis Institute of Technology, Nagpur Campus, Symbiosis International (Deemed University), India*

**Chhabra, Gunjan,** *School of Science and Technology, Swami Rama Himalayan University, India*

**Deoli, Rekha,** *Department of Computer Science and Engineering Graphic Era Hill University, India*

**Dhoundiyal, Shiv Ashish,** *Department of Computer Science and Engineering, Graphic Era Deemed to be University, India*

**Dimri, Sushil Chandra,** *Department of Computer Science and Engineering, Graphic Era Deemed to be University, India*

**Dobhal, Dinesh Chandra,** *Department of Computer Science and Engineering, Graphic Era Deemed to be University, India*

**Gulhane, Monali,** *Symbiosis Institute of Technology, Nagpur Campus, Symbiosis International (Deemed University), India*

**Indu, R.,** *Department of Computer Science and Engineering, Graphic Era Deemed to be University, India*

**Jain, Priyasi,** *Department of Computer Science, Acropolis Institute of Management Studies and Research, India*

**Joshi, Aditya,** *Department of Computer Science and Engineering, Graphic Era Deemed to be University, India*

**Kero, Archana,** *Department of Computer Science, Sai Institute of Paramedical and Allied Sciences, India*

**Kumar, Asheesh,** *GNIOT Institute of Professional Studies, India*

**Kumar, Ashwani,** *Department of Computer Science and Engineering, Shri Ram Group of Colleges, India*

**Kumar, Bhawnesh,** *Department of Computer Science and Engineering, Graphic Era Deemed to be University, India*

**Kushwah, Visheshta,** *Shri Vaishnav Institute of Computer Applications, Shri Vaishnav Vidyapeeth Vishwavidyalaya, India*

**Malik, Preeti,** *Uttar Pradesh Institute of Forensic Science, India*

**Mathur, Sanjay,** *Department of Electronics and Communication Engineering, GBPUA&T, India*

**Mishra, Renu,** *Department of CSE, School of Computing Science and Engineering, Sharda University, India*

**Mittal, Varsha,** *Graphic Era University, India*

**Narwaria, Mamta,** *Department of CSE, School of Computing Science and Engineering, Sharda University, India*

**Negi, Harendra Singh,** *Department of Computer Science and Engineering, Graphic Era Deemed to be University, India*

**Pant, Suman,** *School of Science and Technology, Swami Rama Himalayan University, India*

**Rajput, Ishwari Singh,** *School of Computing, Graphic Era Hill University, India*

**Rakesh, Nitin,** *Symbiosis Institute of Technology, Nagpur Campus, Symbiosis International (Deemed University), India*

**Saxena, Sandeep,** *School of Computer Science and Engineering, IILM University, Greater Noida, India*

**Sharma, Rashmi,** *SVKMs NMIMS University, Mukesh Patel School of Technology Management and Engineering, Shirpur campus, Dhule, Maharashtra, India*

**Singh, Amritansh,** *Symbiosis Institute of Technology, Nagpur Campus, Symbiosis International (Deemed University), India*

**Singh, Murari Kumar,** *Department of CSE, GNIOT, Greater Noida, India*

**Singh, Tinku,** *School of Information and Communication Engineering, Chungbuk National University Cheongju-si, Republic of Korea*

**Upreti, Mitesh,** *Department of Electronics and Communication Engineering, GBPUA&T, India*

**Vinesh,** *IIMT College of Engineering, India*

# List of Figures

# List of Tables

# List of Abbreviations

| | |
|---|---|
| **A4C** | Apical Four-Chamber |
| **AAMI** | Association for the Advancement of Medical Instrumentation's |
| **AF** | Atrial Fibrillation |
| **AI** | Artificial Intelligence |
| **ANN** | Artificial Neural Network |
| **APT** | Advanced Persistent Threats |
| **AUC** | Area Under Curve |
| **AWS** | Amazon Web Services |
| **BaROA** | Bat-Rider Optimization Algorithm |
| **BIH** | Beth Israel Hospital |
| **CHD** | Congenital Heart Disease |
| **CHF** | Chronic Heart Failure |
| **CI** | Computational Intelligence |
| **CNN** | Convolution Neural Network |
| **CPS** | Cyber-Physical Systems |
| **CTI** | Cyber Threat Intelligence |
| **CTS** | Computed Tomography Scan |
| **CV** | Cross Validation |
| **CVD** | Cardiovascular Disease |
| **CWT** | Continuous Wavelet Transform |
| **DB** | Database |
| **DBN** | Deep Belief Network |
| **DCNN** | Deep CNN |
| **DDoS** | Distributed Denial of Service |
| **DL** | Deep Learning |
| **DSC-FL-CNN** | Depth-wise Separable CNN with Focal Loss |
| **DT** | Decision Tree |
| **ECG** | Electrocardiogram |
| **EHR** | Electronic Health Record |
| **EMA** | Emergency Medical Assistance |
| **FDA** | Food and Drug Administration |

| | |
|---|---|
| **FDI** | False Data Injection |
| **FECG** | Fetal Electrocardiogram |
| **FEM** | Finite Element Method |
| **FHD** | Fetal Heart Defects |
| **FLDS** | Feature Learning Detection System |
| **FRCNN** | Faster Region-based CNN |
| **GA** | Genetic Algorithms |
| **GAN** | Generative Adversarial Network |
| **GDPR** | General Data Protection Regulation |
| **GFT** | Graph Fourier Transform |
| **GHz** | Gigaherthz |
| **GRUNET** | Gated Recurrent Unit Network |
| **Gulf SAFE** | Gulf Survey of Atrial Fibrillation Events |
| **HDFS** | Hadoop Distributed File System |
| **HDPM** | Heart Disease Prediction Model |
| **HER** | Electronic Health Record |
| **HFSS** | High Frequency Simulation Software |
| **HIPAA** | Health Insurance Portability and Accountability Act |
| **HLHS** | Hypoplastic Left Heart Syndrome |
| **HRV** | Heart Rate Variability |
| **ICA** | Independent Component Analysis |
| **IDS** | Intrusion Detection Systems |
| **IEM** | Inborn Errors of Metabolism |
| **IoT** | Internet of Things |
| **KNN** | K-Nearest Neighbour |
| **LR** | Logistic Regression |
| **LSTM** | Long short Time Memory |
| **MHRA** | Medical and HealthCare Regulatory Agency |
| **ML** | Machine Learning |
| **MLBHDD** | ML-Based Heart Disease Diagnostics |
| **MOBA** | Multi-Objective Bat Algorithm |
| **MoM** | Method of Moments |
| **MRA** | Magnetic Resonance Angiography |
| **MRCNN** | Mask Region-based CNN |
| **MRHAM** | Multistage Residual Hybrid Attention Module |
| **MRI** | Magnetic Resonance Imaging |
| **NB** | Naïve Bayes |
| **NLP** | Natural language processing |
| **NMF** | Nonnegative Matrix Factorization |
| **PAC** | Probabilistic Analysis and Classification |

| | |
|---|---|
| **PAF** | Paroxysmal Atrial Fibrillation |
| **PCA** | Principal Component Analysis |
| **PCG** | Phonocardiography |
| **PCHF** | Principal Component Heart Failure |
| **PHM** | Public Health Monitoring |
| **PSO** | Particle Swarm Optimization |
| **PTTV** | Pulse Transit Time Variability |
| **QNN** | Quantum Neural Network |
| **$R^2$** | Coefficient of Determination |
| **RBF** | Radial Basis Function |
| **RF** | Random Forest |
| **RNN** | Recurrent Neural Network |
| **RPA** | Robotic Process Automation |
| **SCADA** | Supervisory Control and Data Acquisition |
| **SCM** | Structural Co-Occurrence Matrix |
| **SMOTE** | Synthetic Minority Oversampling Technique |
| **SSIM** | Structural Similarity Index Measure |
| **SVD** | Singular Value Decomposition |
| **SVM** | Support Vector Machine |
| **TA** | Temporal abstraction |
| **TAR** | Temporal Association Rules |
| **WHO** | World Health Organization |
| **XAI** | Explainable Artificial Intelligence |
| **XGBoost** | Extreme Gradient Boosting |
| **RMSE** | Root Mean Square Error |

# 1

# Artificial Intelligence in Healthcare: Revolutionizing Medical Innovations

**Priyasi Jain[1], Visheshta Kushwah[2], Rashmi Sharma[3] and Piyush Chauhan[4]**

[1]Department of Computer Science, Acropolis Institute of Management Studies and Research, India
[2]Shri Vaishnav Institute of Computer Applications, Shri Vaishnav Vidyapeeth Vishwavidyalaya, India
[3]SVKMs NMIMS University, Mukesh Patel School of Technology Management and Engineering, Maharashtra, India
[4]Symbiosis Institute of Technology, Nagpur Campus, Symbiosis International (Deemed University), India

**Email:** jpriyasipj@gmail.com; visheshta2510@gmail.com; rashmi.nov30@gmail.com; shbichauhan@gmail.com

## Abstract

Artificial intelligence in healthcare is introducing the modern trends to have emerged, for better diagnostics and personalizing treatments, along with improving management processes. The artificial intelligence will be put to work sifting through digital health record data, medical images, genomics, and patient-worn device streams to improve clinical decision-making, and hence also the care patients receive, as well as operational efficiency. What was powerful about this fact, though, is that advanced machine learning algorithms coupled with deep learning capabilities were able to see patterns in complex data sets, thereby putting healthcare into a world of predictability and precision. In diagnostics, artificial intelligence is beneficial in identifying diseases such as cancer or neuronal/cardiac disorders like apoplexy at a very early stage for prompt intervention, leading to improved patient outcomes (more relevant where it applies in radiology and pathology). It also



DOI: 10.1201/9788743812739-1

helps in the field of personalized medicine, where treatment can be tailored to everyone based on their genetics or lifestyle, and medical history, for better outcomes with fewer side effects. Aside from clinical, artificial intelligence greatly optimizes hospital operations by performing scheduling, billing, and inventory management functions while freeing health professionals to interact directly with patients. Wearable devices connected to the artificial intelligence algorithm allow a person to monitor his or her health in real-time and, by implication, promote preventive care. However, it would bring ethical and legal questions toward data privacy and other matters relating to the transparency of decision-making processes. Data privacy and the transparency of the decision-making process call for bold regulation. Future promises of change, yet are still being heard. Let artificial intelligence be utilized to pave the road for effective, patient-centered healthcare.

## 1.1 Introduction

AI is being used to revolutionize healthcare by improving diagnostics and treatment, patient care, and enhancing administrative efficiencies. It interprets volumes of healthcare data to generate insights to support clinical decision-making using machine learning, natural language processing, and computer vision.

### 1.1.1 Definition of artificial intelligence

Artificial intelligence (AI) is a branch of science and engineering that focuses on creating and studying intelligent agents, or systems that can see, think, and act in ways that typically require human cognitive abilities. AI aims to capture and make progress on the cognitive function by creating machines that can learn from data, adapt to new scenarios, and perform well under complex situations. It uses theories and methods from computer science, cognitive psychology, philosophy of mind, and pragmatics to explain how these processes are carried out. Ultimately, it aims to develop computational methods to automate the modeling of human-like intelligence.

### 1.1.2 Brief introduction to healthcare

Healthcare is the politically correct way to say management, diagnosis, treatment, and prevention of illnesses, considering mental or physical well-being [1, 2]. The answer could be a plethora of support from medications and medical technology, albeit it also extends to operational services provided by medical professionals such as doctors, nurses, and therapists. Healthcare is a new way of treating diseases and managing chronic diseases that extend life, promotes health, and improves the quality of human life. It includes specialized care to treat specific diseases and general primary care that serves as the entry point for most health systems, focusing on good overall health. New growth avenues in healthcare technologies include telemedicine and electronic health records, faster patient care delivery times, etc.

### 1.1.3 Introduction of artificial intelligence in healthcare

How AI is slowly revolutionizing healthcare, from improving diagnosis, to making treatment more personal, and offering a quickened solution to all of this. Medical imaging AI systems analyze CT, MRI, and X-ray images to help radiologists detect issues with stunning precision, enabling quicker, more accurate diagnoses [3]. The application of AI in genomics enables systematic genetic data diagnosis and therapy prescription for more individualized care [4]. Such AI-driven predictive analytics assess real-time patient data to predict the probability of chronic diseases occurring, thus allowing tailored risk management and preventative measures. Drug discovery AI helps to identify signals or new drugs quickly and easily, it can define drug interactions before they occur, and streamline clinical trial design. The administrative and financial capacity is increased by AI. NLP is a key technology that enables the analysis of unstructured EHR data and plays a significant role in bettering clinical decision-making, enhancing documentation accuracy [5].

Furthermore, patients are more successfully engaged by AI-driven tailored health education and behavioral insights, which encourage improved treatment adherence and healthier lifestyle choices [6]. In general, the use of AI in healthcare leads to innovations that improve patient outcomes and transform the healthcare environment [7], in addition to increasing the precision and effectiveness of therapy.

### 1.1.4 Purpose and scope

AI has recently emerged as a transformative tool in healthcare, with significant impacts on improving diagnostic precision, personalizing treatment, and optimizing operational efficiencies [8]. Recent studies underpin the very real abilities of AI in processing large volumes of medical data, including imaging and EHRs, to uncover patterns that may evade human observation and hence help early disease detection and enhance diagnostic precision. Moreover, AI-powered predictive analytics has been crucial in predicting patients' outcomes and health risks enabling proactive health strategies. AI accelerates the pace of identifying potential compounds in drug discovery through emulated biological interactions, thus saving time and reducing cost compared to traditional approaches [9]. Evidence continues to suggest how effective AI is in remote patient monitoring when continuous health surveillance can be facilitated with real-time data from wearables, allowing timely intervention [10]. AI applied in the healthcare field, therefore, is a promising avenue through which personalized medicine may be advanced and healthcare delivery systems streamlined, and one that, at the very least, deserves further testing and confirmation through clinical trials and longitudinal studies.

However, more research is needed on algorithm development, data security, clinical validation, and ethical issues before integrating AI into healthcare. To fully appreciate AI's disruptive potential in healthcare, these issues must be addressed through thorough research and the establishment of industry standards.

## 1.2 Role of Artificial Intelligence in Healthcare

AI is improving several facets of medical sciences and clinical procedures, including health research. To improve the accuracy and speed of diagnosis, medical pictures, or intricate genomic datasets are evaluated. By spotting possible medicinal molecules early on, AI models expedite the drug discovery process and cut down on expenses and time. Personal medicine with AI planning helps in treatments and combining patient data to provide him better and more efficient treatments. This collection of patient data improves AI in monitoring and automates administrative duties, which streamlines operations in care environments. However, applying AI to health research involves several delicate topics, including algorithm bias, data protection, and ethics, all of which can only be resolved with strict validation and legal compliance.

This entails a structured approach to ensure that AI-driven research addresses the most fundamental healthcare challenges. This includes clearly defining the problem being addressed in research by consulting clinical experts on its relevance, followed by gathering healthcare data, such as medical imaging or EHRs. One of the important stages in AI is data preprocessing, ensuring clean, standardized, and unbiased datasets. The treatment of data should be governed by ethical considerations: privacy for patients and compliance with regulatory requirements. AI techniques are selected depending on the problem-posited ML, DL, and NLP [11], then modeled, trained, validated, and tested with appropriate performance metrics. It is an

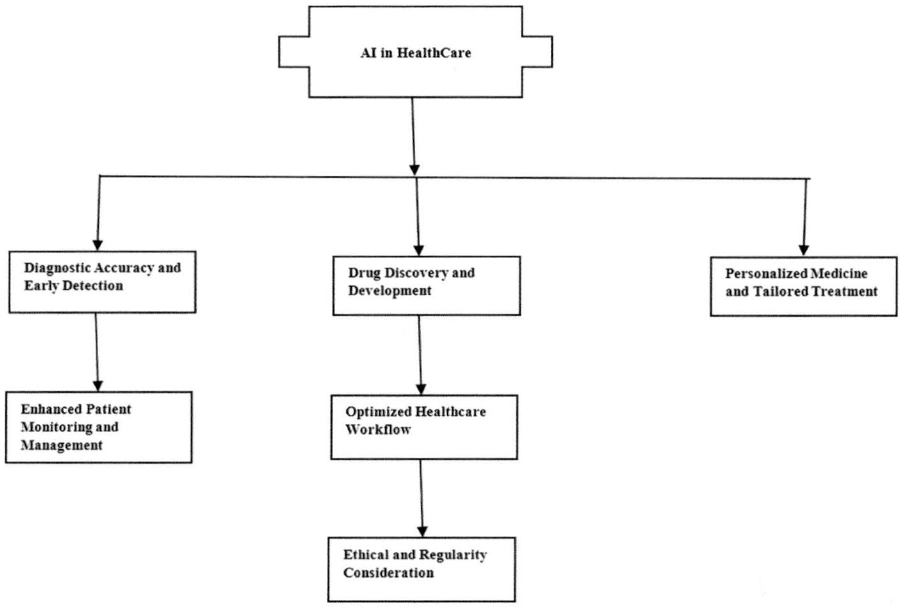


**Figure 1.1** Artificial intelligence (AI) in healthcare.

interdisciplinary approach wherein clinicians make sure that the proposed AI models will fit in with how healthcare is conducted in the real world and within the existing workflows.

Model validation also includes ensuring fairness and targeting biases, which are crucial aspects when one desires to obtain appropriate and generalizable AI solutions. There is a need for the validation of AI models through clinical trials and pilot studies in real-world healthcare settings to prove their effectiveness in diagnosis, treatment, or enhancing operational efficiency. Because regulatory compliance, especially with agencies such as the FDA or EMA, is an important part of the research process in ensuring that AI tools meet the safety and efficacy standards, the last stage of the integration of AI into health practice involves the deployment of AI models into healthcare systems, like hospital records or diagnostic tools, and further monitoring for performance and feedback.

Figure 1.1 shows how AI is used in healthcare, emphasizing how it affects drug discovery, personalized therapy, and diagnostic accuracy. Important topics like early identification, improved patient monitoring, efficient healthcare workflows, and ethical considerations are highlighted in the diagram.

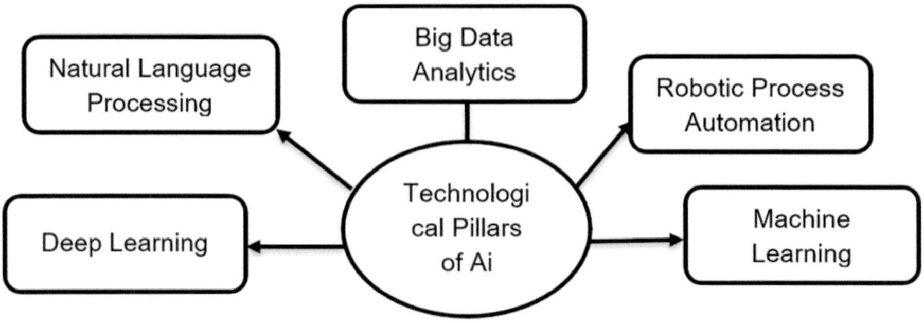


**Figure 1.2** Technological pillars of AI.

## 1.3 Technological Pillars of AI

The significant technological pillars of AI are drivers toward the development and application of systems that belong to the sphere of AI. All these pillars then combine to provide support for advancement and integration across several domains. The application of AI technological pillars in healthcare offers significant advantages that improve the efficacy and efficiency of medical procedures. Figure 1.2 represents the technological pillars of AI.

### 1.3.1 Machine learning and deep learning

A set of algorithms created to analyze [12] and try to comprehend data to predict or use this information is structured and object type called ML, that one can find the transformation brought by AI on how people are working with the health sector. The two main concepts in ML are unsupervised learning, which can find hidden patterns in unlabeled data, using techniques such as clustering or dimensionality reduction while supervised learning uses decision trees and linear regression, etc., to predict outcomes by training models on labeled data. Basic ML techniques include splitting data into training and testing sets to create and evaluate models, as well as using cross-validation to precisely measure performance.

Deep learning (DL), a branch of ML, uses deep neural networks, a multi-layer neural network type, to model complex patterns and representations. Two key concepts in DL are activation functions, which give the model non-linearity, and backpropagation, which uses error gradients to modify network weights. For applications like picture and speech recognition, DL models are particularly good at managing massive amounts of data and automatically extracting characteristics, which makes them very efficient.

Like ML, effective DL model development requires controlling overfitting and utilizing reliable assessment criteria. ML and DL are changing healthcare in diagnostics, treatment, and operational efficiency. The applications of ML range from disease risk analytics, early detection, precision medicine, and medical image analysis to clinical decision support and optimization of resource utilization in hospitals. DL is a more complex subset of ML that has been particularly successful in analyzing complex medical images like image segmentation and object detection, enhancing NLP for information extraction from clinical texts and speech recognition. DL takes genomics further with the analysis of genetic data toward precision medicine and accelerates drug discovery. It will, therefore, enable predictive modeling of disease progression, real-time monitoring of patients, and robotic surgeries with greater precision. Good data quality, consideration of ethics, and collaboration among experts are crucial for effectively using ML and DL in healthcare to drive innovation, improve the delivery of care to patients, and advance medical practices [5, 13, 14].

### 1.3.2 Natural language processing (NLP)

NLP is a subset of AI that focuses on providing meaning to machines to understand, interpret, and talk in a human-like language [15, 16]. NLP combines computational linguistics, ML, and linguistics to provide machines with the ability to process and analyze large amounts of text and speech. The processing of texts, such as tokenization, normalization, part-of-speech tagging to identify grammatical elements in text, named entity recognition to group names and dates, and parsing to understand how sentences are organized are the basic aspects of NLP [1]. Speech recognition, machine translation, and sentiment analysis are other tasks. In addition, question answering uses NLP methods to provide answer-specific facts. Text production generates coherent material automatically. Chatbots, search engines, medical record analysis, and content summarizing are just a few examples of how those functions are used. NLP is therefore crucial for enabling both humans and computers. With NLP, the management and usage of both textual and spoken data are massively enhanced [18]. It simplifies clinical documentation by transcription of spoken notes into structured text, and extraction of relevant information in unstructured medical records so that EHR accuracy and organization improve. It also improves information retrieval by being able to search for medical databases and EHR systems more effectively using semantic search capabilities and supports clinical decisions by retrieving personalized, relevant guidelines and research. NLP supports chatbots and virtual assistants in patient interaction, enabling them to answer patient questions, schedule appointments, and provide health knowledge; symptom checkers can give a preliminary diagnosis based on symptoms described. It also finds applications in NLP for sentiment and emotion analysis in patient feedback and communications regarding satisfaction and monitoring of mental health. It assists in medical research through automated literature reviews and integration of data from disparate sources for comprehensive analyses. In addition, NLP provides support needed for automated medical coding and billing by analyzing clinical documentation and supports predictive analytics through the identification of risk factors and prediction of patient outcomes. In short, NLP improves healthcare through better administrative efficiency, engagement by the patient, clinical decision-making, and research capabilities, thereby improving patients' care and outcomes [16].

### 1.3.3 Robotics process automation

Robotic process automation is a technology based on software robots or "bots" that perform tasks generally executed by humans at high volume, repetitively, and according to rules. RPA emulates human interaction with digital systems to execute tasks such as data entry, filling forms, and transaction processing, all of which can be done without changing existing IT infrastructure. Efficiency is enhanced as the technology reduces manual errors, accelerates the pace of completion of tasks, and assures consistent execution of processes. RPA thus offers very significant cost savings, as the number of workers would be vastly reduced. It is highly scalable and can easily cope with variable workloads. Applications include automation of administrative tasks, customer service processes, and financial operations. Many solutions also involve no-code or low-code platforms; hence, the development and deployment of automation workflows can be done directly by business users. RPA improves productivity, accuracy, and operational efficiency while integrating well with existing systems to release human resources for more value-added and strategic work [19]. RPA can bring efficiency and increased accuracy to healthcare through the automation of repetitive tasks, hence smoothing operations. For example, it simplifies appointment scheduling, data entry, and management of EHRs in administrative and operational areas alone, hence easing administrative burdens and ensuring that patient information is updated. It enhances billing and claims processing by automating the submission of insurance claims and accounting operations that deal with billing. This speeds up reimbursement cycles and reduces errors. In interacting with patients, RPA-driven chatbots and virtual assistants manage routine inquiries and support, improving both patient engagement and onboarding. RPA also integrates clinical workflows, automates activities related to regulatory compliance, and develops timely reports that support data management and decision-making. It further streamlines inventory and supply chain management processes through more effective tracking, ordering, and processing of medical supplies. In a nutshell, RPA enhances operational efficiency in healthcare for better facilitation of patient care and assurance of regulatory compliance through automation of routine tasks and integration among the different processes [20].

### 1.3.4 Big data analytics

Big data analytics refers to the process of analyzing huge and complex data sets for patterns, trends, and insights that could inform decisions based on big data. This process, therefore, encompasses the collection of data, its storage-being able to handle massive volumes of structured, semi-structured, and unstructured information from various sources. Basic techniques involve descriptive analytics for understanding past events, diagnostic analytics to explore causes, predictive analytics to forecast future trends, and prescriptive analytics for recommendations on actions. It allows for such analysis due to advanced tools and technologies including data warehousing systems, big data platforms like Hadoop and Spark, and visualization tools such as Tableau. Applications are found in many industries: in business intelligence, it aids strategic decisions; in healthcare, it develops better patient care; in finance, fraud detection and risk management; in retail, it personalizes marketing and inventory management. It is through these techniques and tools that Big Data Analytics helps organizations reach valuable inferences, implementing operational efficiencies, and leveraging competitive advantages. Big data analytics has a wide range of applications in healthcare: improving patient care, managing operational efficiency, and advancing research. Big data put into practice personalized treatment by analyzing patient data for personalized treatment plans, using predictive analytics for early diagnosis of diseases, and conducting risk management. Operational efficiency in a hospital can be enhanced by optimally managing the resources to reduce costs. The field of clinical research also requires Big data to enable integral data integration or to accelerate drug development. Trend analysis and enhancements to chronic disease management strategies add to the gains in population health management. Data integration opens better diagnostics, accuracy in diagnostic analytics, and even real-time patient monitoring. The experience for patients is further enhanced through feedback insights that optimize their scheduling. Above all, big data analytics drives data-informed decisions, improved patient outcomes, and more effective healthcare delivery [21].

## 1.4 AI-Enhanced Diagnostic Solutions

AI aids in diagnosis in several health domains, including disease prediction, picture medical analysis, and genetic data analysis, leading to effective treatments. The automation of root cause analysis and system failure forecasts is the foundation for progress in the IT domains. AI has numerous uses in the industrial industry, including quality control and predictive maintenance. These systems will continuously learn from new input while integrating data from various sources [22]. Better decisions and outcomes would be attainable because of this integration's notable improvements in accuracy and efficiency regarding the early detection of process issues.

### 1.4.1 Medical imaging and computer vision

Medical imaging automation and analysis are being enhanced by the increasing use of AI-powered tools to analyze images from many modalities, such as CT, MRI, and mammography scans. In one application, these technologies can even be more accurate than a trained radiologist at spotting and diagnosing abnormalities like tumors or lesions. Apart from diagnosis, it also performs image enhancement techniques such as noise reduction and artifact removal. These developments lead to improved scan visibility, enabling the detection of slight irregularities that may otherwise be overlooked.

Automating procedures like segmentation, where AI models identify regions of interest in images (e.g., organs, tumors) for computer vision, are also required. This partitioning is also important for proper therapy planning and disease progression tracking. Anomaly detection also relies on computer vision algorithms to detect patterns corresponding to diabetic retinopathy or fractures. These functions are embedded into radiology workflows, as AI generates preliminary reads to external enterprises and enables prioritizing immediate, life-threatening cases on behalf of a more precise manual review [23].

### 1.4.2 Early disease detection

AI-powered predictive analytics are crucial for early disease detection. These algorithms look at huge swaths of data about patients' genetic information, lifestyle factors, and EHR, among other things, to figure out who is most likely to get certain conditions. AI algorithms are also being developed to be applied for early detection [24], aided by the performance of wearable technologies and remote monitoring. Devices with sensors and AI algorithms can measure blood pressure, pulse rate, and sugar levels over continuous durations. By identifying unusual patterns of data or abnormalities from our baseline health level, these tools could help warn people and healthcare professionals about potential problems before symptoms become severe. This differs from most wearables in that the companies have shown that real-time monitoring can help you prevent chronic illnesses and acute health catastrophes. Advances in technology (from AI and beyond for early disease detection), on the other hand, are changing things by allowing us to spot diseases faster with more precise diagnosis before proceeding or confirming them, improving patient outcomes overall. The integration of AI into medical imaging, wearable technologies (medicine delivered by sensors), gene sequencing, and predictive analytics in the future is going to allow us many more advancements in early detection and preventative treatment.

AI is also changing the diagnosis and treatment of rare diseases that remain undiagnosed for a long time due to a reduced amount of data about their research, and scarce information available on them. With the advanced pattern recognition of AI in genomic data, it can identify rare genetic mutations and establish links between these mutations and specific diseases, cutting down the time taken for diagnosis by a huge margin. Thus, AI analyzes the data in global databases, which may indicate a link between rare diseases and other types of conditions. It helps to show some new and maybe even unthought-out treatments to apply to the patient.

### 1.4.3 Pathology

Through the automation and improvement of medical picture processing, AI and computer vision are profoundly changing pathology. AI algorithms can help in the identification and classification of cellular structures using digital pathology, which entails scanning glass slides into high-resolution digital formats. In oncology, where AI can identify malignant tissues, gauge the aggressiveness of tumors, and forecast patient outcomes based on integrated clinical data, this technology enhances diagnostic accuracy. AI expedites workflows by lowering turnaround times for diagnostics, automating repetitive tasks, and facilitating remote consultations.

AI-powered interactive learning platforms with feedback and simulations are improving pathologists' training. Notwithstanding these advantages, there are still issues with data protection, system validation, and workflow integration. In the future, AI is anticipated to play a bigger role in pathology, providing more advanced diagnostic tools, and improving early illness diagnosis and individualized therapy.

## 1.5 AI Innovation in Custom Medicine

AI is changing the game in drug discovery and development [4], significantly accelerating this whole process and personalizing it toward an individual genetic profile [25]. "Drug discovery has long been associated with delays and over-expenditure, but AI accelerates target identification, compound screening, and drug repurpose." AI could study vast datasets – genomic and proteomic data – to identify drug targets based on predictions for how chemical compounds might interact with biological pathways. It even proposes new chemical structures for potentially tailored treatments to the individual's condition. This is most useful for complex diseases, like cancer, where treatments must be highly specific for individual patients' genetic mutations. More importantly, it identifies subgroups of patients most likely to benefit from a particular therapy, eliminating those that will not help, and this predicts the risk of failing the clinical trial. Hence, treatment will be more successful and tailored according to the requirements of the patient. Personalized dosing and treatment plans will be based on changing genetic makeup, as well as accessing real-time data of health and environmental impacts. This is rarely considered, but AI can include genetic information, including the way a patient metabolizes specific drugs, to ensure that dosage is personalized. Dynamic dosing allows for constant modification since the AI allows wearable technologies to monitor real-time physiological responses, such as blood glucose or heart rate, in a patient. Moreover, AI prevents drug adverse interactions through the examination of how one's genetic profile and the current situation may cause several types of medication to interact. Treatments with AI are more accurate [26] and safer, while being better tailored to the needs of every patient by factoring in lifestyle and environmental variables. In regard to rare diseases, this is helpful when finding off-label drug uses or new drug combinations, which brings up customized treatment options for patients. It would also enable the identification of suitable candidates for clinical trials using the genetic and clinical data available with respect to rare diseases to accelerate research and development within underserved areas.

## 1.6 AI-Driven Efficiency in Healthcare Management

AI systems can automate nearly all the routine administrative tasks like scheduling, patient registration, and billing. NLP tools can manage appointment bookings and automatically create a data entry process, which would reduce the administration staff workload and decrease human error. This can be used for the organization of EHR, so the medical record is accurate, up-to-date, and accessible to healthcare providers, saving much time and money. This will minimize human error, reduce diagnostic time, facilitate faster medical interventions, and produce better patient outcomes within the appropriate use of healthcare resources [18].

### 1.6.1 Electronic health record management with AI

Automatic entry of data to EHRs can be said to be one of the primary areas through which AI is shaping EHRs; from the clinical side, it enhances accuracy and makes better decisions. Among key benefits lies the automation of capturing through AI, particularly through NLP, which transcribes physician-patient communications and updates medical records almost in real time. This has the benefit for healthcare providers, reducing administrative burden, allowing them to spend more time caring for patients with accuracy in data [27]. In addition to automation, AI can be utilized for providing clinical decision support, which analyzes data pertinent to patient care and provides a diagnosis or potential treatment [28, 29]. This, in turn, enables fast and accurate medical decisions. Further, an AI system can utilize predictive capabilities to analyze trends in the data of patients and predict possible health risks in advance to allow early intervention. These predictive analytics, combined with AI's ability to detect and correct errors in the EHRs, leave patient records accurate and intact, thereby reducing the case of misdiagnoses and inappropriate treatments. AI personalizes the treatment of the patient [30] by using historical data, life factors, and genetics to adapt approaches to individual treatment for better care and avoid trial and error. Further, AI workflow can be optimized to give preference to data-containing information crucial to the patients, like lab abnormalities, and in organizing the information so that access can be quick, which also improves efficiency within the healthcare system. We can easily understand the steps of HER with the help of Figure 1.3.

### 1.6.2 AI for optimizing hospital operations

The introduction of AI in the optimization of health operations has brought a change in the approach of managing resource utilization, patient care, and

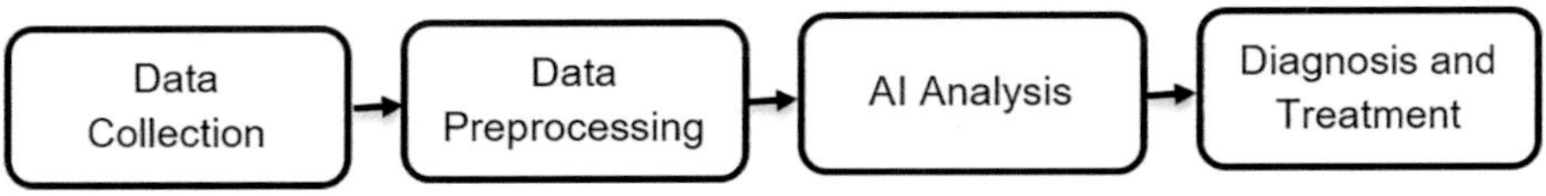


**Figure 1.3** EHR with AI.

administrative processes within healthcare systems, thereby making the system more productive and responsive [31]. The primary benefit of AI in healthcare is the prediction of demand for patients and the ability to help with resource allocation. AI-based predictive analytics [24] can examine historic data and predict which patients will require admission for treatment, what their healthcare needs will be, and so forth, and it can ensure proper allocation of staff, equipment, and beds in hospitals and clinics. That way, it minimizes overstaffing and understaffing, decreases patient waiting times, and utilizes resources appropriately. It further enhances supply chain management by making an estimate of the demand for drugs, equipment, and other inputs to the hospitals; it automates inventory tracking; and ensures shortages or overstocking do not arise, which saves some of the costs of the healthcare organizations and works against operational breakdowns. The third role that AI plays is in improving [32, 33] patient flow through identifying bottlenecks that can be found in hospital workflows and tailoring scheduling and patient movement for more optimal services. AI can predict busy periods by analyzing the data of patients, and real-time adjustments in the operations by healthcare administrators will help in smooth transitions between different stages of care and the reduction of waiting times. Improvements have been made in telemedicine and remote patient monitoring with the inclusion of AI. Routine activities will be handled by AI-based tools, such as virtual assistants or chatbots, to enable people to assess their symptoms, schedule an appointment, or reply to queries for patients. It will leave more health personnel to deal with complex situations. At the same time, the AI-based remote monitoring systems would monitor the patients' vital signs and health data in real-time, thus facilitating better management of large patient volumes, generally for chronic disease management. On the administrative side, AI automates most of the routine tasks in billing, coding, and insurance claims processing. It eliminates human error and hastens processes that otherwise take a lot of time. AI can be used to analyze patient data for predicting and preventing hospital readmissions based on the identification of who is at risk, thereby allowing patient-centered follow-up care. This lightens the burden of resources in the hospitals and improves patients' outcomes. AI facilitates data-driven decision-making for healthcare administrators, as it provides actionable insights into operational performance as well as trends that reflect patient care. This helps administrators make better decisions in terms of resources, human capital, and long-term planning to improve patient care. Additionally, it reduces waste and fraud by identifying inefficiencies in the way operations are carried out, inconsistencies in billing, and anomalies in insurance claims. Ultimately, AI significantly enhances healthcare operations, making the operations far more efficient and cost-effective while focusing on enhancing the scope of patient care.

### 1.6.3 Virtual health assistants and chatbots

A virtual health assistant and a chat board will provide a holistic health management support system. The virtual health assistant is one powered by AI as a personal health guide. It can include what symptoms users are experiencing and let them get probable diagnoses, reminders about taking their medications, and other personal health tips [34, 35]. The virtual assistant can schedule appointments and can be integrated to allow connection with wearables for real-time monitoring of the patients' vital signs. Personal interaction with the virtual assistant keeps the users on track with health goals [36]. However, the chat board invokes community. The users can discuss health topics among themselves, share personal experiences, and seek advice from others who have similar conditions. Specialized forums on diabetes or mental health, for instance, help make the individuals feel safe while trying to connect and look for support. Expert Q&A allows professionals in healthcare to share insights and questions and to widen the knowledge of the community. Together, the virtual health assistant and the chat board [37] form an interactive system. For example, there is the possibility that people might discuss their symptoms on the chat board as the assistant advises them in relation to their needs or organizes appointments in respect to their conditions. It could also be that the assistant's records of progress sort out related discussions on health improvement and encourage people. This integration would therefore not only improve the management of an individual's health but also form a supporting community in which users find knowledge from each other and empowerment on respective journeys toward health. Working together, these tools serve as a dynamic platform for holistic wellness.

## 1.7 Revolutionizing Public Health Monitoring with AI

Revamping the field of public health monitoring AI will help revolutionize PHM to the point that it promises to transform the very way we understand, address, and deal with health issues at the local and global levels. For instance, AI will be able to streamline the real-time collection of data coming from all sorts of sources, including social media and devices worn by people, EHRs, and many more. Instead, ML algorithms can more rapidly work through a massive pool of health data to determine if certain areas are displaying troubling signs or early evidence that an outbreak has arrived in the country. Predictive analytics-based AI also analyzes the previous medical records to predict health habits and outbreaks. This is helpful because it allows public health organizations to take pre-emptive actions to boost community preparation and ensure the population will be more resilient when putting resources or vaccines into place. Supporting these initiatives, advanced AI-based surveillance that automatically identifies outlying health data as computers quickly identify an odd trend in symptoms that could have gone unnoticed. Otherwise, this could include immediate notifications for spikes in on-line symptom reports. Another major positive impact of AI in public health is the development of targeted health campaigns. With the classification of demographic data and a description of health behavior, AI can help in designing public health activities specific to communities that will most likely respond favorably to them. More participation happens, and the effectiveness of any health promotion is improved when messages are based on the unique needs and preferences of populations. For instance, AI can allocate resources optimally, ensuring that healthcare facilities are placed in areas where they are needed most. Predictive models can help predict hospital admissions based on the causative relationships between variables that may cause hospital admissions, thus effectively managing staff and resource utilization in hospitals. AI-based decision support systems are one of the crucial agents, providing insights based on complex datasets available to the health authority, which is guiding policy making as well as intervention strategies [22].

## 1.8 Ethical and Legal Challenges of AI in Healthcare

The integration of AI in healthcare research introduces several legal and ethical challenges that relate to the sensitivity of health data and the dynamic, technologically paced landscape. One of the major issues it raises is the issue of privacy and data security [26]. AI systems work on massive amounts of personal health information (PHI), making them susceptible to breaches. Legally, respecting such regulations as HIPAA in the United States or GDPR in the EU would impose severe constraints on handling health data. Ethically, there is an important requirement that patient consent and privacy must be respected [38]; otherwise, improper use or de-identification of data can still pose a risk of patient re-identification. Bias and fairness are crucially important concerns as well. Those AI algorithms, in turn, have the potential to unintentionally reflect and amplify all the biases present in the data being used to train them. Consequently, biased outcomes might favor discriminatory decisions against an underserved population. Legally speaking, biased healthcare decisions would likely fall under the mandates of anti-discrimination laws like the Civil Rights Act or ADA. From an ethical standpoint, healthcare AI must be aligned with fairness, demonstrating that it does not worsen existing inequalities nor bring new forms of bias that could disadvantage certain demographic groups. Another issue within AI healthcare research is the lack of transparency and accountability. Most AI systems, especially DL-based ones, can be described as "black boxes." Their decision-making processes are not easily interpretable. Legally, this complicates assigning responsibility in cases where AI has made an error. Whether it is the role of the developer, the healthcare provider, or the AI system remains ambiguous. Ethically, transparency in healthcare is important because any decision that would affect patient care should be trusted and understood by both the patients and providers. The issue of informed consent is more complex with AI-driven research. AI typically retrieves or relies on secondary data that repurposes patient information without direct consent for each use, thereby placing them at risk in terms of the law: most data protection laws require consent from patients for specific purposes. Ethically speaking, using data without patient consent undermines patient autonomy and creates potential avenues for exploitation if patients do not know about their data usage. Liability is yet another open issue in the landscape of AI in healthcare. As the number of patient harms associated with decisions by AI-driven algorithms increases, it is increasingly complex to resolve liability for those involved in the making and implementation of those decisions. Present frameworks for healthcare liability in product liability or medical malpractice suits do not recognize AI as a causation factor. Ethically, it is critical that patient harm issues are addressed as soon as possible while considering guidelines that ensure the providers will not be liable for actions driven by AI systems and do not occur due to human error or some other cause that negatively affects the patient. Finally, issues involving intellectual property (IP) and ownership arise in AI healthcare research. AI systems, datasets, and innovations often blur the lines of ownership. From a legal perspective, a lack of clarity on intellectual property rights could easily lead to disputes between researchers, institutions, or indeed commercial entities. Ethical issues demand a balance between the greater benefit of public versus commercial interests and to ensure that these AI-driven health innovations have greater good without monopolies exploiting the technology [39].

### 1.8.1 Limitations of ethical challenges

1. Prejudice and equal treatment: Disparities in healthcare outcomes may result from AI algorithms that inherit biases from training data.

2. Accountability and understanding: A lot of AI algorithms, particularly deep learning models, operate as "black boxes," which makes it challenging to justify choices.
3. Patient independent thinking: AI suggestions may jeopardize informed decision-making and the doctor-patient relationship.
4. Data security and privacy: Managing private patient information increases the possibility of security lapses and abuse.
5. Transparency and obligation: It's unclear who bears the blame when AI provides a false diagnostic or suggestion.
6. Clinical faith and recognition: Without conclusive proof, medical practitioners may be hesitant to embrace AI-driven judgments.
7. Dehumanization of Care: If AI is used excessively, patient interactions may become less human.

## 1.9 Regulatory Frameworks and Guidelines

The most important regulatory frameworks and guidelines on AI in healthcare should ensure that the use of AI within a clinical space is safe, ethical, transparent, and effective. They then discuss some of the challenges, including data privacy, patient safety, bias, and accountability, while promoting innovation in healthcare. Some of the aspects include that regulatory bodies like the U.S. should ensure testing and validation of rigorous AI systems [35]. The FDA, or the European Medicines Agency (EMA), classifies AI medical devices and applications based on risk. Data privacy is the biggest issue, considering that laws such as HIPAA in the U.S. or GDPR in Europe mandatorily govern the accountability of AI data systems to safeguard patient information with permission. Along with interpretability and transparency, it should be evident for which reasons an AI is concluding something, especially when it is critical to diagnosis or treatment recommendations. It is significant that regulatory bodies have emphasized the implementation of interpretable AI models. Furthermore, AI systems should address the problem of bias and fairness to ensure healthcare equity across race, gender, or socioeconomic status. In regulatory advice, such guidelines would require monitoring of the actual deployment in the field and ongoing post-market surveillance that describes actual safety concerns or failures as they could arise once the AI is deployed in the operational setting. Ethical considerations also prevail; AI should augment rather than replace human judgment in the delivery of patient-centered care. Such regulatory bodies as the FDA, EMA, and WHO oversee the process, and, more importantly, a global approach puts emphasis on matters of safety, equity, and ethical applications. Of course, these frameworks are not aimed at burdening patients but to add to responsible and innovative uses of AI in healthcare [40]. Various aspects are mentioned in Table 1.1 [3].

**Table 1.1** Aspects of AI in healthcare.

| **Aspect** | **Decision** | **Key regulatory bodies** | **Significance** |
|---|---|---|---|
| Patient safety and effectiveness FDA, 2021; EMA, 2022; MHRA, 2023) | AI systems must be tested for safety and performance before use in healthcare. Regulatory bodies categorize AI based on the risk to patients. | FDA (U.S.), EMA (Europe), MHRA (UK) | Prevents patient harm by guaranteeing the safety, dependability, and clinical efficacy of AI-driven healthcare technologies. |
| Data privacy and security (HIPAA, 1996; GDPR, 2016; Health Canada, 2022) | AI must comply with data protection regulations, ensuring secure handling of patient data and obtaining patient consent. | HIPAA (U.S.), GDPR (EU), Health Canada | Ensures moral and legal compliance by safeguarding patient privacy and preventing the improper use of private medical information. |
| Transparency and explainable (FDA, 2023; WHO, 2021) | AI decision-making processes must be transparent and explainable, allowing healthcare providers and patients to understand how conclusions are reached. | FDA, EMA, WHO | Increases confidence in AI systems by guaranteeing that medical personnel can decipher and verify AI-generated judgments. |
| Bias and fairness (WHO, 2021; FDA, 2022) | AI developers must address and minimize bias in their algorithms to ensure equitable healthcare for all demographic groups. | FDA, EMA, WHO | Makes sure AI-based judgments are impartial and fair, which improves outcomes for all patients and lessens health inequities. |

*(Continued)*

**Table 1.1** *Continued.*

| Aspect | Decision | Key regulatory bodies | Significance |
|---|---|---|---|
| Continuous monitoring and post-market surveillance (FDA, 2023; EMA, 2022) | AI systems require ongoing monitoring after deployment to ensure continued performance and safety in real-world applications. | FDA (U.S.), EMA (Europe), Health Canada | Identifies dangers or real-world failures in AI applications, enabling prompt updates and enhancements. |
| Ethical consideration (WHO, 2021; National Ethics Committees, Various) | AI systems must respect patient autonomy, uphold ethical principles, and support human judgment rather than replace it in healthcare decision-making. | WHO, National Ethics Committees (various countries) | Ensures that AI avoids becoming overly dependent on automation while upholding ethical medical procedures, human dignity, and doctor-patient relationships. |
| Key regulatory bodies (FDA, 2023; WHO, 2022) | Various regulatory organizations oversee the safe implementation and monitoring of AI in healthcare, providing global and national guidelines. | FDA (U.S.), EMA (Europe), WHO (global), MHRA (UK), Health Canada | Creates uniform rules and regulations to guarantee the ethical, safe, and efficient use of AI in healthcare. |

## 1.10 The Future of AI in Healthcare

Improved patient outcomes, reduced complexity, and increased accessibility and personalization will be seen to make the future of AI in healthcare revolutionize the healthcare sector. For personalized medicine, AI would be an essential component [41]: streamlining treatments proven effective or ineffective for treatment based on a patient's genetic makeup, medical history, and lifestyle. Its ability to analyze vast datasets will also enable early detection and disease prediction, especially for complex diseases such as cancer and cardiovascular diseases. AI can spot patterns [42] in medical scans and predict outbreaks of diseases, so interventions can be made earlier, and management of population health can be improved. It will also speed drug discoveries significantly, reducing the time and cost of bringing new treatments to market. Telemedicine, enabled through AI-driven virtual assistants for health, will boost access to healthcare, offering immediate support and remote consultations primarily in underserved areas. Robotic surgery will increase accuracy, while AI will take care of all the administrative tasks, making such health systems even more efficient. In mental health, AI will diagnose disorders early and offer treatment through apps and tools targeted to personal needs. AI with wearable devices will help in monitoring patients from a distance, detecting health-related issues in real-time. AI will also improve the management of EHRs: the automation of data entry and extraction from insights, thus giving healthcare providers much more time to take care of patients. The integration of AI will also be accompanied by further development in the ethical and regulatory frameworks that ensure the safe, transparent, and equitable use of AI in healthcare. Ultimately, AI will enhance healthcare delivery [26] in ways that make the processes more specific, efficient, and accessible while maintaining the crucial role of human healthcare providers.

The AI-powered healthcare system (Figure 1.4) acts as a central hub that connects patients, providers, researchers, regulators, and databases. It enables improved diagnostics, efficient data management, and compliance with ethical standards while leveraging AI-driven insights for enhanced healthcare delivery.

1. Deep learning in radiology: AI-driven image recognition models (CNNs) will be able to identify abnormalities in CT, MRI, and X-ray images with unprecedented precision.
2. Multi-modal AI models: These models combine genomic, imaging, and textual data (such as IBM Watson and Google's Med-PaLM) to provide a comprehensive diagnosis.
3. Merging of liquid biopsy with AI: AI-assisted diagnosis of cancer using non-intrusive blood tests that examine circulating tumor DNA (ctDNA).
4. Federated learning for secure AI training – AI models will train across decentralized hospital networks without exposing sensitive patient data.
5. Real-time haptic feedback: During operations, surgeons will receive comprehensive resistance and texture data from AI-enhanced touch sensors.
6. Synthetic biology and AI: CRISPR gene editing powered by AI will allow for precise genomic changes to treat genetic illnesses.

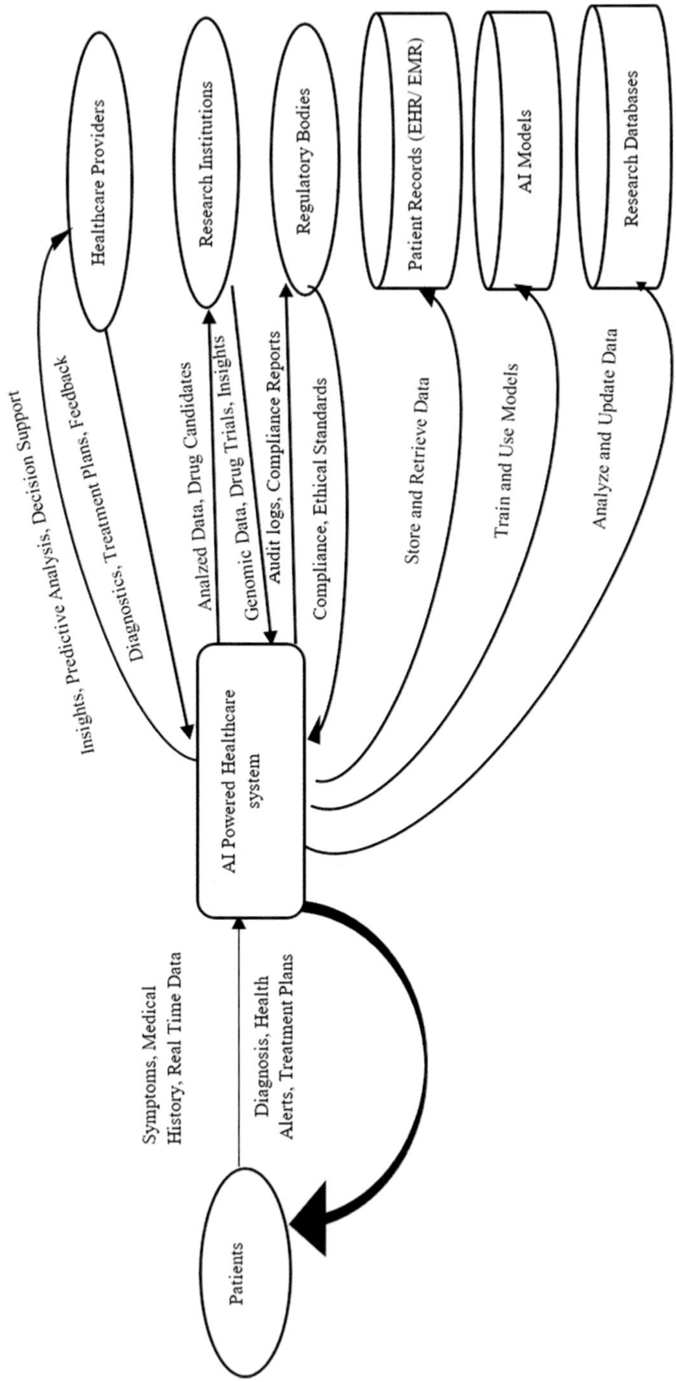


**Figure 1.4** Artificial intelligence (AI)-powered healthcare system.

## 1.11 Conclusion

Artificial intelligence (AI) is set to transform healthcare delivery, moving from a one-size-fits-all model to a more personalized, efficient, and accessible healthcare system, fundamentally changing how diseases are detected, diagnosed, and treated. AI can mine massive volumes of medical records and detect diseases like cancer, Parkinson's, or Alzheimer's even before the human body manifests any symptoms, ensuring timely treatment and making diagnostics more reliable. These also create opportunities for personalized treatment and care tailored to the history and genetic profile of each patient. Apart from clinical applications, AI increases efficiency in drug discovery, playing a vital role in the acceleration of drug discovery and commercialization processes for pharmaceutical companies, improving administrative tasks, and rationalizing hospital workflows and resource allocation, decreasing the cost of running a healthcare facility, and increasing overall business efficiency. By facilitating telemedicine and remote patient monitoring, it enhances healthcare accessibility, particularly for chronic disease management – and along similar lines, AI-powered robotics and imaging systems enable greater surgical precision, as well as better diagnostic capabilities. Meanwhile, the use of AI in public health surveillance can monitor an outbreak and forecast the spread of diseases to assist in prevention. But harnessing these benefits entail overcoming some crucial challenges, particularly around robust data protection, the safeguarding of patient privacy, transparency in and the mitigation of bias in algorithms to ensure that care is equitable and fair. Just as research models are evolving, regulatory frameworks will also need to provide more clarity in this balancing act of innovation versus patient safety. To sum it up, AI will evolve into an integral part of contemporary healthcare, pushing it toward a more preventive, inclusive, and patient-centric paradigm with improved outcomes, seamless patient journeys, and a more resilient and equitable system as a whole.

## References

[1] Zhou, B., Yang, G., Shi, Z., & Ma, S. (2024). Natural language processing for smart healthcare. *IEEE Reviews in Biomedical Engineering, 17*, 4–18. https://doi.org/10.1109/RBME.2022.3210270

[2] Ahmed, I., Jeon, G., & Piccialli, F. (2022). From artificial intelligence to explainable artificial intelligence in Industry 4.0: A survey on what, how, and where. *IEEE Transactions on Industrial Informatics, 18*(8), 5031–5040. https://doi.org/10.1109/TII.2022.3161011

[3] Bharati, S., Mondal, M. R. H., & Podder, P. (2024). A review on explainable artificial intelligence for healthcare: Why, how, and when? *IEEE Transactions on Artificial Intelligence, 5*(4), 1429–1442. https://doi.org/10.1109/TAI.2023.3266418

[4] Topol, E.J. (2019). High-performance medicine: the convergence of human and artificial intelligence. *Nature Medicine, 25*(1), 44–56. https://doi.org/10.1038/s41591-018-0300-7

[5] Huang, C., Zhang, Z., Mao, B., & Yao, X. (2023). An overview of artificial intelligence ethics. *IEEE Transactions on Artificial Intelligence, 4*(4), 799–819. https://doi.org/10.1109/TAI.2022.3194503.

[6] Virmani, N., Singh, R. K., Agarwal, V., & Aktas, E. (2024). Artificial intelligence applications for responsive healthcare supply chains: A decision-making framework. *IEEE Transactions on Engineering Management, 71*, 8591–8605. https://doi.org/10.1109/TEM.2024.3370377.

[7] World Health Organization (WHO). (2022). *AI in healthcare: Global regulatory frameworks*. WHO.

[8] Zahlan, A., Ranjan, R. P., & Hayes, D. (2023). *Artificial intelligence innovation in healthcare: Literature review, exploratory analysis, and future research. Technology in Society, 74*, 102321. https://doi.org/10.1016/j.techsoc.2023.102321

[9] Qureshi, R., et al. (2023). Artificial intelligence and biosensors in healthcare and its clinical relevance: A review. *IEEE Access, 11*, 61600–61620. https://doi.org/10.1109/ACCESS.2023.3285596

[10] Umbare, R., Patil, R., Mukund, T., Rathod, A., & Rajurkar, P. (2024). *A smart healthcare application using artificial intelligence (AI) and machine learning (ML). In Proceedings of the 5th International Conference on Image Processing and Capsule Networks (ICIPCN)*, Dhulikhel, Nepal, 204-210. https://doi.org/10.1109/ICIPCN63822.2024.00041

[11] Surve, R., Purohit, T., Joseph, R., & Shaikh, P. (2023). *HealthCare chatbot using machine learning and NLP. In Proceedings of the 6th International Conference on Advances in Science and Technology (ICAST)*, Mumbai, India, 411-416. https://doi.org/10.1109/ICAST59062.2023.10455027

[12] Ahsan, M.M., Luna, S.A., & Siddique, Z. (2022). Machine-learning-based disease diagnosis: a comprehensive review. *Healthcare, 10*(3), 541. https://doi.org/10.3390/healthcare10030541

[13] Chakraborty, C., Bhattacharya, M., Pal, S., & Lee, S. S. (2023). From machine learning to deep learning: Advances in the recent data-driven

paradigm shift in medicine and healthcare. *Current Research in Biotechnology*, 100164.

[14] Rathore, D. K., & Mannepalli, P. K. (2021). A review of machine learning techniques and applications for health care. In *Proceedings of the 2021 International Conference on Advances in Technology, Management & Education (ICATME)* (pp. 4–8). IEEE. https://doi.org/10.1109/ICATME50232.2021.9732761

[15] Khurana, D., Koli, A., Khatter, K., & Singh, S. (2022). Natural language processing: State of the art, current trends, and challenges. *Multimedia Tools and Applications, 82*(6), 8171–8207. https://doi.org/10.1007/s11042-022-13428-4

[16] World Health Organization. (1948). *Constitution of the World Health Organization*. https://www.who.int/about/governance/constitution.

[17] Dubey, K., Bhowmik, M., Pawar, A., Patil, M. K., Deshpande, P. A., & Khartad, S. S. (2023). Enhancing operational efficiency in healthcare with AI-powered management. *In 2023 International Conference on Artificial Intelligence for Innovations in Healthcare Industries (ICAIIHI)* (pp. 1–7). IEEE. https://doi.org/10.1109/ICAIIHI57871.2023.10488953

[18] Yang, G., et al. (2020). *Homecare robotic systems for Healthcare 4.0: Visions and enabling technologies. IEEE Journal of Biomedical and Health Informatics, 24*(9), 2535-2549. https://doi.org/10.1109/ JBHI.2020.2990529

[19] Kaur, J. (2023). Robotic process automation in the healthcare sector. *E3S Web of Conferences, 391*, 10008. https://doi.org/10.1051/e3sconf/202339101008

[20] Ahmed, A., Xi, R., Hou, M., Shah, S. A., & Hameed, S. (2023). Harnessing big data analytics for healthcare: A comprehensive review of frameworks, implications, applications, and impacts. *IEEE Access, 11*, 112891–112928. https://doi.org/10.1109/ACCESS.2023.3323574

[21] Alowais, S.A., Alghamdi, S.S., Alsuhebany, N., et al. (2023). Revolutionizing healthcare: the role of artificial intelligence in clinical practice. *BMC Medical Education, 23*, 689. https://doi.org/10.1186/s12909-023-04698-z

[22] Javaid, M., Haleem, A., Singh, R. P., & Ahmed, M. (2024). Computer vision to enhance healthcare domain: An overview of features, implementation, and opportunities. *Intelligent Pharmacy, 2024*, 1-10. https://doi.org/10.1016/j.ipha.2024.05.007

[23] Premanand Ghadekar, A. D. (2024). *Early disease detection and prediction using AI technologies: Approaches, future outlook, mitigation*

*strategies, and synthesis of systematic reviews*. International Journal of Intelligent Systems and Applications in Engineering, 12(3), 1434–1445. https://www.ijisae.org/index.php/IJISAE/article/view/5536

[24] Blasiak, A., Khong, J., & Kee, T. (2020). CURATE.AI: Optimizing personalized medicine with artificial intelligence. *SLAS Technology, 25*(2), 95–105. https://doi.org/10.1177/2472630319890316

[25] Santamato, V., Tricase, C., Faccilongo, N., Iacoviello, M., & Marengo, A. (2024). *Exploring the impact of artificial intelligence on healthcare management: A combined systematic review and machine-learning approach. Applied Sciences, 14*(22), 10144. https://doi.org/10.3390/app142210144

[26] Liang, J., He, Y., Xie, J., Fan, X., Liu, Y., Wen, Q., Shen, D., Xu, J., Gu, S., & Lei, J. (2023). Mining electronic health records using artificial intelligence: Bibliometric and content analyses for current research status and product conversion. *Journal of Biomedical Informatics, 146*, 104480. https://doi.org/10.1016/j.jbi.2023.104480

[27] Elkefi, S., & Asan, O. (2022). *Exploring doctors' satisfaction with electronic health records and their features: A nationwide survey study. In Proceedings of the 2022 IEEE International Symposium on Systems Engineering (ISSE)*, Vienna, Austria, 1-5. https://doi.org/10.1109/ISSE54508.2022.10005506

[28] Okediran, O. O., Sijuade, A. A., Wahab, W. B., & Oladimeji, A. I. (2020). *A framework for a cloud-based electronic health records system for developing countries. In Proceedings of the 2020 International Conference on Electrical, Communication, and Computer Engineering (ICECCE)*, Istanbul, Turkey, 1-5. https://doi.org/10.1109/ICECCE49384.2020.9179276

[29] Khan, I. A., et al. (2024). Big data analytics model using artificial intelligence (AI) and 6G technologies for healthcare. *IEEE Access, 12*, 97924–97937. https://doi.org/10.1109/ACCESS.2024.3427333

[30] Maleki Varnosfaderani, S., & Forouzanfar, M. (2024). The role of AI in hospitals and clinics: Transforming healthcare in the 21st century. *Bioengineering (Basel)*, 11(4), 337. https://doi.org/10.3390/bioengineering11040337

[31] Badawy, M., Ramadan, N., & Hefny, H. A. (2023). Healthcare predictive analytics using machine learning and deep learning techniques: A survey. *Journal of Electrical Systems and Information Technology*. https://doi.org/10.1186/s43067-023-00108-y

[32] Sai, S., Gaur, A., Sai, R., Chamola, V., Guizani, M., & Rodrigues, J. J. P. C. (2024). Generative AI for transformative healthcare:

A comprehensive study of emerging models, applications, case studies, and limitations. *IEEE Access, 12*, 31078–31106. https://doi.org/10.1109/ACCESS.2024.3367715

[33] Jiang, F., Jiang, Y., Zhi, H., Dong, Y., Li, H., Ma, S., Wang, Y., Dong, Q., Shen, H., & Wang, Y. (2017). Artificial intelligence in healthcare: past, present and future. *Stroke and Vascular Neurology, 2*(4), 230–243. https://doi.org/10.1136/svn-2017-000101

[34] Pradhan, B., Das, S., Roy, D. S., Routray, S., Benedetto, F., & Jhaveri, R. H. (2023). An AI-assisted smart healthcare system using 5G communication. *IEEE Access, 11*, 108339–108355. https://doi.org/10.1109/ACCESS.2023.3317174

[35] Punugoti, R., Duggar, R., Dhargalkar, R. R., & Bhati, N. (2023). *Intelligent healthcare: Using NLP and ML to power chatbots for improved assistance. In Proceedings of the 2023 International Conference on IoT, Communication and Automation Technology (ICICAT)*, Gorakhpur, India, 1-6. https://doi.org/10.1109/ICICAT57735.2023.10263708

[36] Bhukya, B., Sheshikala, M., & Bhukya, B. (2023). *NLP based topic modeling for healthcare: Analyzing patient reviews to improve quality of care and access to services. In Proceedings of the 2023 International Conference on Emerging Techniques in Computational Intelligence (ICETCI)*, Hyderabad, India, 7-12. https://doi.org/10.1109/ICETCI58599.2023.10330957

[37] Olawade, D. B., David-Olawade, A. C., Wada, O. Z., Asaolu, A. J., Adereni, T., & Ling, J. (2024). *Artificial intelligence in healthcare delivery: Prospects and pitfalls. Journal of Medicine, Surgery, and Public Health*, 3, 100108.

[38] Bataineh, A., Mushtaha, A., Abu-AlSondos, I., Aldulaimi, S., Al-Ekir, E., & Abdeldayem, M. (2024). Ethical & legal concerns of artificial intelligence in the healthcare sector. In *2024 International Conference on Emerging Trends in Smart Information Systems (ICETSIS)*. IEEE. https://doi.org/10.1109/ICETSIS61505.2024.10459438

[39] Mennella, C., Maniscalco, U., De Pietro, G., & Esposito, M. (2024). Ethical and regulatory challenges of AI technologies in healthcare: A narrative review. *Heliyon, 10*(4), e26297. https://doi.org/10.1016/j.heliyon.2024.e26297

[40] Paranjape, K., Schinkel, M., & Nanayakkara, P. (2020). Short keynote paper: Mainstreaming personalized healthcare–transforming healthcare through new era of artificial intelligence. *IEEE Journal of Biomedical and Health Informatics, 24*(7).

[41] Maher, C., Singh, B., Wylde, A., & Chastin, S. (2024). Virtual health assistants: A grand challenge in health communications and behavior change. *Frontiers in Digital Health, 6*, 1418695. https://doi.org/10.3389/fdgth.2024.1418695

# 2

# Beyond Human Insight: How AI Is Crafting the Next Era of Healthcare Innovation

**Preeti Malik**[1] **and Varsha Mittal**[2]

[1]Uttar Pradesh State Institute of Forensic Science, India
[2]Graphic Era University, India

**Email:** preetishivach2009@gmail.com; var.aadi@gmail.com

## Abstract

In medical diagnostics, artificial intelligence and machine learning have revolutionary potential, especially for atrial fibrillation or other heart-related diseases. However, the broad use and efficacy of artificial intelligence-based diagnostic systems are hampered by several issues. These include the need for better training datasets to get over data scarcity challenges, energy-efficient and precise early detection models, and explainable artificial intelligence to address transparency concerns. Opportunities for improved diagnostics arise from the investigation of alternative sensing technologies, such as wearable technology, in conjunction with the incorporation of big data analytics. Standardization and interoperability are still essential for developing scalable, financially feasible solutions for a range of demographics. According to new research, artificial intelligence can detect cardiovascular disease with the same accuracy as a cardiologist. This proves how artificial intelligence and machine learning approaches can be applied to heart disease, potentially preventing hundreds of thousands of unnecessary deaths worldwide each year. This study primarily aims to review recent artificial intelligence approaches for the detection of heart diseases, mainly arrhythmia, heart failure, and congenital heart disease. This study emphasizes the potential and difficulties of using artificial intelligence and machine learning for atrial fibrillation



DOI: 10.1201/9788743812739-2

detection, highlighting the necessity of multidisciplinary cooperation and creative methods to deal with these urgent problems.

## 2.1 Introduction

At the center of life itself lies the human heart, a relentless pump that beats more than three billion times in an average lifespan. Nevertheless, despite decades of developments in the healthcare sector, the disease related to the heart remains the major reason for death worldwide. In 2023, the global population reached 8 billion! Of these, approximately 620 million people worldwide are living with CVDs [1]. In 2021 alone, CVD accounted for 20.5 million deaths, comprising approximately one-third of all global deaths. Figure 2.1 shows death trends from 1990 to 2019. Each year, around 60 million individuals develop a CVD. CVDs encompass various conditions impacting the heart and blood vessels. These are rheumatic heart disease, coronary artery disease, cerebrovascular disease, and some other associated illnesses. CVD can lead to:

- Arrhythmia (abnormal heartbeat)
- Blockages in the heart's blood vessels

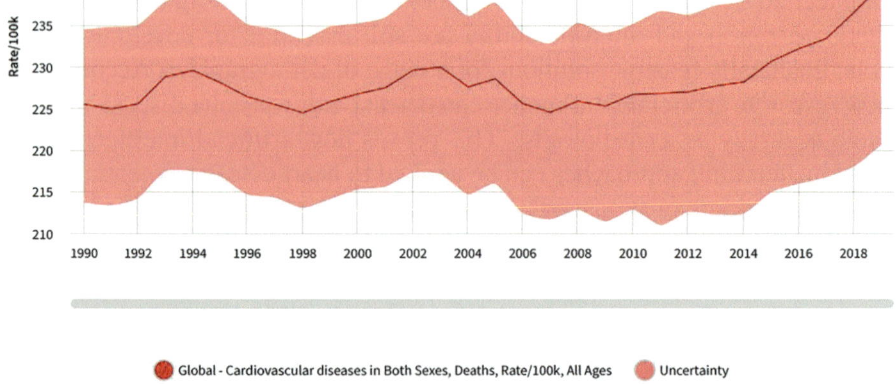


**Figure 2.1** Trends in cardiovascular disease, 1990–2019 (Source: https://world-heart-federation.org/world-heart-observatory/trends/).

- Atrial fibrillation (abnormal heartbeat in the upper parts of the heart, preventing normal blood flow)
- CHD (heart issues present at birth, affecting different parts of the heart)
- Heart failure (abnormal pumping of the blood by the heart)

Any of these conditions can increase the risk of stroke or blood clots. Computer-aided methods are more effective than traditional manual practices for detecting various heart diseases. In the battle against this unrelenting foe, a quiet revolution is taking place today, one that is powered by data and algorithms rather than stethoscopes and scalpels. The idea of AI is no longer limited to sci-fi fantasies; it is now a potent tool for the detection, management, and even prevention of cardiac disease. The issue of how far we can push the limits of what robots can accomplish to save human life emerges when ML and medicine meet. According to new research, AI can detect CVD with the same accuracy as a cardiologist. This proves how AI and ML approaches can be applied to heart disease, potentially preventing hundreds of thousands of unnecessary deaths worldwide each year.

This study primarily aims to review recent AI approaches for the detection of heart diseases, mainly arrhythmia, heart failure, and CHD [2]. Additionally, the high cost of current diagnostic facilities makes them inaccessible to many people. CVD is diagnosed by ECGs; however, it is time and effort-consuming to visually recognize long-term ECG irregularities. The rise of ML applications in the medical field has led to the development of MLBHDD systems by medical professionals and researchers. These systems are adaptable and reasonably priced [3] [4] [5].

## 2.2 Background

ML trains computers to use past experiences to solve problems, gaining popularity due to affordable computing power and memory [6]. This allows for processing and analyzing large datasets to uncover hidden insights and correlations. ML's intelligence is based on algorithms that enable abstraction and accurate judgments. DL, a more sophisticated branch of ML, automatically extracts, analyzes, and understands useful information from raw data by mimicking human thought processes. It employs neural data and automatic feature engineering, enhancing accuracy and performance. ML and DL's success relies on classification algorithms, but DL is increasingly replacing ML due to superior feature extraction from nonlinear hidden layers. ML relies on handcrafted features, while DL's automatic learning results in better classification performance.

### 2.2.1 Classifiers

Data analyzed under ML and DL must undergo preprocessing and feature extraction processes. Preprocessing means converting raw data into information. Features, which are specific characteristics of the data, must be discriminative and non-redundant to reduce training time and prevent overfitting. Various methods exist for feature extraction, particularly in heart disease detection, as outlined in Section 4. Once features are extracted, the data can be labeled using classifiers – algorithms that categorize the data. Common classifiers in ML include SVM, RF, LR, DT, NB, and KNN. Unlike ML's step-by-step approach, DL creates a network similar to a biological neural network to handle the entire process. DL employs many nonlinear computational unit layers, each of which converts input into more abstract forms for the layer after it. DL classifiers include RNN, CNN, Boltzmann machines, autoencoders, and DBN.

### 2.2.2 Performance metrics

The evaluation of ML and DL detection systems for accurately classifying CVD relies on performance parameters like accuracy, precision, recall, specificity, AUC, and F1-score. Each performance metric provides different insights into the model's effectiveness. For instance, a model might excel in accuracy but perform poorly in specificity. Here is a table (see Table 2.1) summarizing the performance metrics for ML algorithms along with their formulas:

**Table 2.1** Performance metrics for ML algorithms.

| Metric and formula | Description |
|---|---|
| $Accuracy = \frac{TP + TN}{TP + TN + FP + FN}$ | Ratio of correct predictions to the total number of predictions. |
| $Specificity = \frac{TN}{TN + FP}$ | Ratio of true negative to all actual negatives. |
| $AUC$ | AUC measures the overall ability of the model to discriminate between classes. |
| $Precision = \frac{TP}{TP + FP}$ | Ratio of true positive to all positive predictions. |
| $Recall = \frac{TP}{TP + FN}$ | Ratio of true positive to all actual positives. |
| $F1\ score = \frac{2.Precision.Recall}{Precision + Recall}$ | It is the harmonic mean of precision and recall. |

### 2.2.3 Data for heart disease detection

One of the key factors in detecting different types of heart diseases is the source of data. Many symptoms, including gender, age, pulse rate, chest discomfort, blood pressure, sugar level, and resting ECG, can be used to predict CVDs [7] [8]. Analyzing the data in the healthcare sector helps in disease prediction, improves diagnosis, analyzes symptoms, provides appropriate medications, enhances the quality of care, reduces costs, extends life spans, and lowers the death rate among heart patients. For instance, ECG assists in screening irregular heartbeats and strokes by using embedded sensors placed on the chest to monitor the patient's heartbeat. Detailed clinical data play a crucial role in heart disease prediction, aiding experts in making informed decisions. This data can come in the form of electronic health records, wearable device readings, patient surveys, and so on.

### 2.2.4 Contribution

The primary contributions of this article are outlined below:

- Recent studies have been reviewed on heart diseases, including arrhythmia, CHF, and heart failure, utilizing ML and DL to pinpoint the most effective detection methods.
- A concise summary of the heart disease DBs most commonly referenced in the analyzed studies has been included.
- A summary of the key conclusions drawn from the examined studies has also been provided, along with a discussion of several unresolved problems and possible avenues for further study.

### 2.2.5 Paper organization

The rest of the content is organized as follows: Section 2.3 is committed to the different DBs. Section 2.4 provides a literature review on three different heart diseases. Section 2.5 demonstrates existing review articles in this field. Section 2.6 summarizes the key findings from the reviewed articles and discusses various open research issues and potential future directions. Section 2.7 concludes the paper. For a quick overview, the organization of this article is illustrated in Figure 2.2.

## 2.3 DBs for Heart Disease

**UC Irvine Repository** [9]: It is a repository of 655 datasets that are open-source related to different medical fields and helps the community of machine

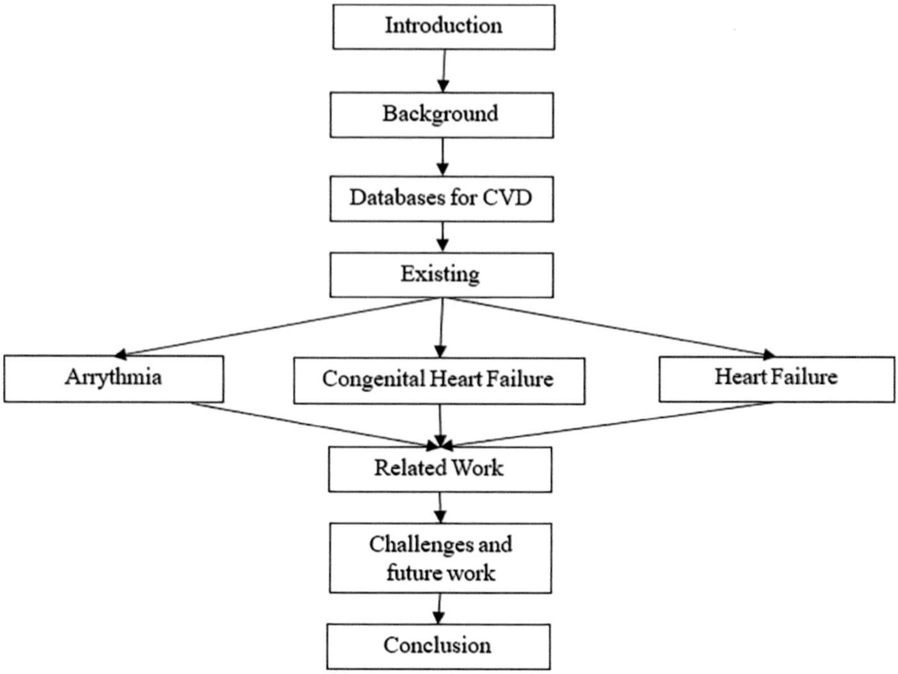


**Figure 2.2** Organization of the article.

learning. The heart disease dataset of this repository consists of four DBs, namely: Switzerland, Cleveland, VA, Long Beach, and Hungary. Although this DB has 76 properties, all published studies use only a subset of 14. This dataset is multivariate and is mainly used for a classification task. The dataset contains 303 instances and has missing values, so it needs to be pre-processed before it can be used for analysis. All the features are either categorical, integer, or real values.

**Gulf Survey of AF Events** [10]: The Gulf SAFE is a potential, exploratory registry for AF patients. 2043 consecutive AF patients were registered at 23 hospitals across six Middle Eastern Gulf nations from October 2009 to June 2010. Data were obtained using a standard form and input online. This data includes demographic information of the patient, AF type, medical history, and therapy. The 12-month follow-up is planned to conclude in July 2011. The average age was 57, and out of them, more than 50% were male. Hypertension is the most common concurrent condition, which was found in 1072 patients. The dataset reveals that the patients had a history of coronary artery disease was 28%, 30% diabetes, and rheumatic heart disease in 16%. Nine percent of patients had a history of stroke, and four percent had a history of transient ischemic episodes. 37% of patients experienced their first attack of AF, the most common kind of AF, while 19% had lone AF.

**MIT-BIH AF 2000** [11]: This DB contains 25 long-term ECG recordings from human individuals with AF (mainly paroxysmal). Among all, 23 records embrace two ECG signals: 00735 records represent rhythm (.atr) files, whereas 03665 represents unaudited beat as .qrs file. Two ECG signals, each collected at 250 samples per second with 12-bit resolution over a range of ±10 mV, are included in each of the 10-h recordings. The Beth Israel Deaconess Medical Centre, previously the BIH in Boston, used ambulatory ECG recorders to make the first analogue recordings.

**MIT-BIH Arrhythmia DB** [12]: BIH Arrhythmia Labs examined 47 individuals, and their data is kept in the Arrhythmia DB. This data is in the form of two-channel ambulatory ECG recordings of half-hour clips, and these are 48 in number. A mixed cohort of patients, comprising roughly 60% outpatients and 40% inpatients, at Boston's BIH provided a total of 4000 24-h ambulatory ECG recordings. The recordings were converted to digital format.

**PAF Prediction Challenge DB** [11]: This two-channel ECG recording DB was developed for the Computers in Cardiology Challenge 2001, a nationwide competition aimed at creating automated techniques for PAF prediction. The DB is partitioned into learning and testing datasets. All the records of the type n* and P* are categorized as the learning set, and the records of the type t* are taken as the test set. There are fifty record sets in the learning set. The 50 record sets are derived from 48 separate subjects, and each of the four records in a record set is an excerpt of a lengthier continuous ECG recording of that individual subject. Every record has multiple files attached to it. The digitized ECGs are stored in *.dat files, using 16 bits per sample with the least significant byte first in each pair.

**Intra-cardiac Atrial Fibrillation DB** [13]: Eight patients with AF have ECG recordings from their right atria included in this DB. In four different cardiac areas, a decapolar catheter with 2-5-2mm spacing (7mm between bipoles) was inserted. Three surface ECG leads and five bipolar signals were obtained in each location. At 1 kHz, the data were digitized. The labels for the catheter bipoles are CS12, CS34,..., and CS90 (distal to proximal).

**AF Termination Challenge DB** [14]: This DB is created with the objective to develop automated algorithms for forecasting impulsive termination of AF by Computers in Cardiology Challenge 2004, an open competition. This collection of two-channel ECG recordings has been produced for that purpose. Two test sets are created with 30 and 20 records, as well as a learning set with 30 records. Every record comprises two ECG signals, each recorded at a rate of 128 samples per second, and represents a one-minute segment of AF. The segments were taken out of ECG records that lasted between 20 and 24 hours. An automatic detector creates a set of comments for each record, labelling all detected beats – including ectopic beats – as normal. These annotations might have a few minor mistakes in them because they haven't been audited. The task is to determine whether test set records demonstrate self-terminating AF out of the one-minute excerpts of two-lead long-term ECG recordings that reveal continuous or self-terminating AF.

**2017 PhysioNet** [15]: The 2017 PhysioNet/CinC Challenge aims to inspire the creation of processes capable of determining whether a single short ECG recording (lasting 30 to 60 s) displays normal sinus rhythm, AF, another type of rhythm, or is too noisy to classify. Cardiac arrhythmias can be categorized based on origin, rate, mechanism, AV conduction, and duration.

**Record AF** [16]: Record AF is the world's first 1-year observational longitudinal study. It is the Registry on Cardiac Rhythm disORDers monitoring the control of AF. It was carried out in 532 locations in 21 countries across America, Asia, and Europe. The main goals were to evaluate the clinical results and treatment efficacy of rhythm and rate-control techniques prospectively. Both the patient baseline data and the trial design are disclosed. Out of 5814, 5604 AF patients were registered as eligible for evaluation.

**Cardiovascular Disease Dataset** [17]: Three well-known heart disease datasets were combined to create this curated dataset. The first dataset was gathered via Kaggle. It has 70,000 records with 11 distinct variables. These were obtained at the time of the patient's informational interview and medical checkup. Thirteen similar traits are shared by 303 and 293 instances, respectively, in the second and third datasets. Its curation is based on three datasets: Heart-related Data (Kaggle Dataset), Cleveland (Repository for UCI ML), (UCI ML Repository) in Hungarian.

Table 2.2 highlights the key aspects of each dataset, including their description, the number of instances or records, key features, data types, and any additional notes that provide context or specific details about the dataset.

**Table 2.2** Datasets with their features.

| Dataset | Instances/ records | Key features | Data types | Remarks |
|---|---|---|---|---|
| **UC Irvine Repository** | 303 | 76 properties (14 commonly used); categorical, integer, or real values. | Multivariate | Used for classification; contains missing values requiring pre-processing. |
| **Gulf Survey of AF Events** | 2043 | Patient demographics, type of AF, medical history, treatment, and ER visit outcome. | Various | Data collected using a standardized case report form and input online. |
| **MIT-BIH AF 2000** | 25 recordings | ECG signals, rhythm (.atr) files, and unaudited beat (.qrs) files. | ECG signals | Collected at 250 samples/ second with 12-bit resolution; manual and automatic annotations. |
| **MIT-BIH Arrhythmia DB** | 48 snippets | Digitized ECG recordings at 360 samples/ second/channel with 11-bit resolution. | ECG signals | Part of a larger set of 4000 24-h ambulatory ECG recordings. |
| **PAF Prediction Challenge DB** | 50 record sets (learning) | ECG recordings in learning and testing datasets; machine-generated annotation files. | ECG signals | Contains multiple files per record; includes 30-minute and 5-minute continuation records. |
| **Intracardiac Atrial Fibrillation DB** | 8 patients, 4 records each | ECG leads and bipolar signals, sampled at 1 kHz with 14-bit resolution. | Intracardiac and ECG signals | Recordings from four different cardiac areas for each patient. |

(*Continued*)

**Table 2.2** *Continued.*

| Dataset | Instances/ records | Key features | Data types | Remarks |
|---|---|---|---|---|
| **AF Termination Challenge DB** | Learning: 30 records, Testing: 50 records | ECG signals, automatic QRS annotations. | ECG signals | Contains one-minute segments of AF from long-term recordings; used for classification of self-terminating AF. |
| **2017 PhysioNet** | Not specified | Classifies cardiac arrhythmias based on origin, rate, mechanism, AV conduction, and duration. | ECG signals | Challenge focused on classifying rhythms from short recordings. |
| **Record AF** | 5,814 (5,604 eligible) | Patient baseline data, treatment efficacy, clinical outcomes. | Various | Conducted across 532 locations in 21 countries. |
| **Cardiovascular Disease Dataset** | 70,000 (Kaggle), 303 (UCI), 293 (UCI) | Patient data collected during interviews and checkups; 11 variables (Kaggle), 13 traits (UCI datasets). | Various | Combines data from Kaggle, Cleveland, and Hungarian datasets. |

## 2.4 Strategies for Heart Disease Prediction

This section is dedicated to the existing approaches proposed by various researchers that make use of AI/ML/DL approaches for disease prediction or any kind of assistance in the healthcare industry. Our focus is on arrhythmia, Congenital disease, and heart failure.

### 2.4.1 Strategies for addressing arrhythmia

Liu et al. in [18] developed a novel feature extraction method that integrates bi-spectrum analysis with the 2D-GFT. They restructured high-order matrices from the bi-spectrum into datasets that could be transformed into the eigenvalue spectrum domain using the GFT, enabling the identification of features from the eigenvalues' statistical characteristics. These spectral features were then calculated to create a feature vector. The group used the MIT-BIH DB to categorize different arrhythmic heartbeats following the AAMI guidelines. The results of the experiment demonstrated that the suggested model, which combines bi-spectrum analysis and 2D-GFT, achieved a classification accuracy of 96.2% utilizing an SVM with a RBF and CV.

Marinho et al. evaluate various feature extraction techniques [19]. Notably, this study marks the first utilization of SCM for ECG signal feature extraction. Results from the evaluation of some well-known classifiers were contrasted with those from six traditional feature extraction techniques. The study adheres strictly to the AAMI protocol and utilizes the MIT-BIH Arrhythmia DB to ensure robust clinical analysis outcomes. Findings indicate a 2% higher confidence level in detecting heart dysrhythmias compared to existing literature methods. Moreover, the proposed system demonstrates a 1.3% improvement in accuracy over the best-reported approach to date, with significantly faster operation. In summary, employing HOS for arrhythmia classification proves clinically reliable, achieving an accuracy of 94.3%.

Zou et al. introduce a new attribute referred to as a segment label to improve heartbeat classification accuracy [20]. Generated by a CNN, this feature captures contextual information around each heartbeat. The MIT-BIH DB was used for training and evaluating a random forest classifier that combined the segment label with conventional characteristics. With an overall accuracy of 0.96 and noteworthy precision and sensitivity for differentiating between pulse types, the results demonstrate competitive performance.

The approach suggested by Ji and Zhu automates feature selection, bypassing manual selection challenges [21]. The GRUNET DL model is introduced, utilizing CNNs to extract ECG signal features, enhancing their utilization. To capture temporal dependencies, GRUs are integrated. The method achieves remarkable results: 99.47% accuracy on the European ST-T DB, 98.76% precision, and 97.92% recall on the MIT-BIH DB, with 100% accuracy. These findings highlight the method's superiority, reducing manual intervention and improving classification accuracy. Khan et al. introduce a DL-based system that employs CNNs to categorize ECG signals and use PhysioNet MIT-BIH Arrhythmia DB [22]. Five heartbeat types were successfully classified in the test dataset. A ten-fold CV is used to assess the classifier's performance. The findings show an average of 99.06% specificity, 92.86% precision, 92.41% sensitivity, and 98.63% accuracy. Furthermore, the average kappa and F1-score are 95.5% and 92.63%, respectively. The paper emphasizes how ResNet achieves better results than other 1-D CNNs.

Using unbalanced ECG datasets, a DSC-FL-CNN approach was proposed for automatically classifying arrhythmia by Lu et al. [23]. The classification performance was much improved by the focused loss, especially for arrhythmias with fewer data. The DSC-FL-CNN model also decreased the number of parameters needed. To train the model on the MIT-BIH DB, which was then assessed for 17 distinct arrhythmia classifications. The suggested approach demonstrated a significant increase in arrhythmia classification, achieving an F1-score of 0.79 when compared with the contemporary techniques currently in use. Merdjanovska and Rashkovska used three other public datasets with comparable arrhythmia groups to demonstrate a CNN's cross-DB generalization capabilities [24]. The CNN was previously effective in identifying arrhythmias on the MIT-BIH DB. To find out how well-established models can categorize arrhythmias in a broader group of people who weren't in the neural network's training set, the focus is on practical assessment techniques, particularly inter-patient evaluation. Results indicate that the cross-DB generalization performance declines if the conditions (such as lead position) under which the measurements were conducted in the other DBs differ from those in the MIT-BIH DB.

The aim of the research presented by Gaddam et al. is to implement a DL-based algorithm for classifying different cardiac arrhythmias [25]. Initially, CWT translates 1-D ECG impulses into 2-D scalogram pictures. To assess the suggested method, four ECG waveform categories were chosen from the PhysioNet MIT-BIH DBs. Main aim is to create a transferable DL system to automatically classify the four distinct heart conditions. The final findings showed that a noteworthy accuracy of 95.67% was attained by the 2-D scalogram pictures trained with a deep CNN utilizing the transfer learning approach (AlexNet). Consequently, it is clear that the technique described here functions as a training set for an efficient automated heart disease detection tool network.

Atal and Singh utilized an optimization-based deep CNN for arrhythmia classification. This includes the BaROA, which combines the ROA with the MOBA [26]. To represent distinct ECG aspects, wave and Gabor features are first taken from the ECG signals. To distinguish between arrhythmia and non-arrhythmia circumstances, these signals are then fed into a BaROA-based DCNN classifier. The MIT-BIH DB is used to test the approach, and performance is rated according to sensitivity, specificity, and accuracy. The results show that the network's training set has accuracy, specificity, and sensitivity values of 93.19%, 95%, and 93.98%, respectively.

A CNN with long short-term memory was incorporated to create a classification method for arrhythmias to identify eight ECG signals [27]. The MIT-BIH DB provided the ECG data. The two main parts of the experimental strategy were converting 1-D signals into 2-D grayscale pictures for the model's input data and then utilizing the combined model to identify and categorize the input data. Surprisingly, this technique eliminates the necessity for independent ECG signal noise filtering or feature extraction. High classification performance was demonstrated by the experimental findings, with the network's training set achieving 99.01% accuracy, 99.57% specificity, and 97.67% sensitivity, respectively.

### 2.4.2 Strategies for addressing congenital heart disease

To handle fetal ultrasound pictures and identify heart wall abnormalities using many objects, the study by Nurmaini et al. [28] uses MRCNN. It is the first instance segmentation-based medical application for septal defect identification. In contrast to alternative methods of object detection like FRCNN, training is twice as quick when using the MRCNN architecture with ResNet50 as the backbone and a 0.0001 learning rate. Mean average precision shows a strong association between ground truth and anticipated septal flaws. The results show that the suggested MRCNN model performs well in the multiclass detection of heart chambers.

To accurately segment seven structures in the A4C view, a unique end-to-end cascaded U-net model known as CU-net with SSIM loss was proposed by Xu et al. [29]. Two U-nets with branch supervision and between-net links make up the predict-refine CU-net architecture. The technique successfully collects global information and maintains distinct tissue borders by including the SSIM loss function. Firstly, the CU-net with two branch supervisions enhances tissue boundary clarity and mitigates the gradient vanishing problem associated with increased network depth. Secondly, between-net connections transmit prior information from shallow layers to deeper layers, refining segmentation results. Lastly, leveraging SSIM loss preserves fine-grained structural information and delineates clear boundaries. Experimental results on A4C view datasets demonstrate that the proposed method achieves 85.6% Dice Similarity Coefficient, 3.311 Hausdorff Distance, and 92.9% Pixel Accuracy.

The aim of the study by Rahman et al. [30] was to utilize ML techniques to identify morphological and temporal abnormalities in cardiac ultrasound videos of fetuses with HLHS. Echocardiography videos were collected at various gestational stages from fetal HLHS patients and pre-processed for training. Five different DL CNN-LSTM models and a novel stacking CNN-LSTM model were trained using these videos. Results indicate that the stacking CNN-LSTM model, developed using MobileNetv2-LSTM, ResNet18-LSTM, and GoogleNet-LSTM models with different pre-trained encoders, is highly effective in distinguishing HLHS patients from healthy ones using ultrasound videos. In terms of cross-sectional heart chamber dimensions and flow hemodynamics, the model correctly distinguishes between healthy and HLHS human fetal hearts throughout the gestational development phases. With video-wise classification accuracy, precision, sensitivity, F1 score, and specificity of 90.5%, 92.5%, 92.5%, 92.5%, and 85%, respectively, the stacking CNN-LSTM model performed better than previous pre-trained CNN-LSTM models. Likewise, the model's accuracy, precision, sensitivity, F1 score, and specificity for subject-wise classification using ultrasound videos were 90.5%, 92.5%, 92.5%, 92.5%, and 85%, respectively.

Ziani et al. focuses on detecting FECG signals from a single-channel abdominal lead using a combination of CNN and advanced mathematical methods, including ICA, SVD, and NMF for dimension reduction [31]. The extremely unbalanced frequency of the fetus' heart rate in relation to the mother's is identified in terms of energy by using time-scale representation. To optimize the actual FECG signal, which is improved using the SVD-ICA procedure, different parts of the ECG can be separated and used as inputs to the CNN model. The findings and graphs that relate accuracy values to loss values, displaying a 94% performance level, indicate how effective the CNN model is.

Gong et al. in [32] introduce a novel model called DGACNN, which demonstrates superior performance in recognizing FHD, and achieve a rate of 85%. DANomaly and GACNN (Wgan-GP and CNN) are the two primary components of the DGACNN architecture. To further enhance FHD recognition, a small number of DANomaly-screened annotated video slices can be used for data amplification. According to experimental findings, DGACNN outperforms other cutting-edge networks in identifying FHD by 1% to 20%.

An intelligent FLDS for fetal cardiac views is proposed by Qiao et al. to detect the four chambers [33]. The FLDS integrates a MRHAM to learn powerful and robust features, enabling accurate localization of the four chambers in fetal cardiac views. Numerous tests demonstrate that our suggested FLDS outperforms the state-of-the-art, attaining an F1 score of 0.944, a precision of 0.919, and a recall of 0.971. Xiao et al. describe the construction of a new pediatric heart sound dataset, which was gathered and annotated in actual clinical settings by qualified medical professionals [34]. This dataset's abundance and annotations facilitate the development of a DL-based computer-aided system for diagnosing pediatric CHDs. Building on this dataset, two distinct 1-D CNN-based models are used to create a lightweight pediatric CHD detection system with an emphasis on parameter efficiency and feature reuse. The results showed that the created system performs better than the state-of-the-art CNN-based techniques and can be used as a monitoring system for the early detection of CHDs due to its notably low parameter consumption. The best accuracy of 96.21% is attained by the suggested dense block-based approach.

Using widely accessible and reasonably priced PCG, a DL-based attention transformer model was created for the detection of heart murmurs brought on by CHD by Alkhodari et al. [35]. Experts gathered and analyzed PCG recordings from 942 young individuals at four primary auscultation locations as either missing, present, or unknown murmurs. A 10-fold CV was used to validate performance, achieving 90.23% accuracy and 72.42% sensitivity. When tested on unseen data, the accuracy of distinguishing between the presence and absence of murmurs was 76.10%. When predicting the occurrence of murmurs in newborns, children, and teenagers, the model showed accuracies of 70%, 88%, and 86%, respectively.

The primary objective of the work done by Ungureanu et al. is to create an Intelligent Decision Support System that can detect the presence or absence of previously taught significant characteristics using 2D video files of cardiac sweeps taken during a regular first-trimester FECG by [36]. The research seeks to address the key restraints of existing methods for automating FECG, which typically focus on second-trimester scans and solely rely on DL for diagnosing CHD.

### 2.4.3 Strategies for addressing heart failure

The study by Ning et al. [37] presents a model for automatically detecting CHF, which employs a mix of DL algorithms consisting of RNN and CNN. This model categorizes CHF signals and normal sinus heart rate signals using an ECG. The algorithm analyses 5-minute ECG signals with 99.93% accuracy, 99.85% sensitivity, and 100% specificity.

Qadri et al. suggest applying machine learning techniques to detect cardiac failure [38]. To build the models, it uses a dataset that includes 1025 patient records. The eight most noticeable features are chosen to improve performance using a brand-new PCHF feature engineering technique. Various ML algorithms are analyzed and compared. The suggested DT approach requires only 0.005 s of runtime computations while achieving 100% accuracy. Every learning model undergoes cross-validation using a 10-fold approach to verify performance. Zou et al. present a novel ECG signal classification model using 2000 intervals between R-peaks [39]. The proposed model combines encoders, decoders, and intermediate units to preserve low-level details and extract complex information related to pathology. Some enhancements are also done by the author. To address diverse sample collection, data is decentralized across different clients. Experimental results on the data from PhysioBank demonstrate better performance than other approaches.

A new method for precise CHF detection is presented by Prabhakararao et al. [40]. It successfully retrieves high-level attentive representations and captures temporal dynamics. Before multilayered RNNs with residual connections extracting high-level data hierarchically, an RNN layer encodes sequential dynamics from raw ECG beats. The network's modeling ability is improved by the residual connections, which enable efficient gradient propagation. In order to improve CHF identification, an attention module then finds hidden vectors that correlate to diagnostically significant ECG features. With ECG signals from publicly accessible datasets, the suggested approach achieves a remarkable 98.57% beat-level diagnosis accuracy and a nearly 100% 24-h record-level diagnostic accuracy. By improving model simplicity and detection effectiveness, the study of learnt attention weights shows a focus on clinically significant ECG signals that characterize CHF, hence increasing research in CHF analysis.

Wang et al. proposed a classifier for detecting CHF using short-term HRV and DNNs. With blindfold validation, high accuracies were achieved across different DBs, and RR segment lengths [41]. The superior performance is attributed to DL features enabling reliable signal abstraction without human intervention. Combining DL features with expert features enhances performance. However, the study has limitations, notably disregarding data imbalance issues.

In ref. [42], Zhao et al. explore the potential enhancement in heart failure detection by simultaneously analyzing HRV and PTTV in both healthy persons and heart failure patients. Forty subjects from each group were enrolled. HRV and PTTV analysis were conducted using various indices. Significant differences were observed in HRV indices between healthy and heart patients, though only MEAN in PTTV significantly decreased in heart failure patients. This research underscores the importance of PTTV analysis in detecting heart failure. Ishaq et al. in [43] propose an ML-based technique for predicting heart patients' survival. Class imbalance is reduced by the SMOTE method, and RF is also used for feature extraction. Comparative analysis on both full and selected feature sets from the dataset demonstrates the effectiveness of tree-based models. SMOTE enhances the performance of tree-based classifiers.

Gjoreski et al. introduced a hybrid method for detecting CHF from audio recordings of PCG, combining classical ML and end-to-end DL [44]. Evaluation on our dataset and six publicly available PhysioNet datasets demonstrated superior performance compared to baseline methods. Additionally, we extended the study to personalize models for detecting different CHF phases. By identifying 15 features with varying distributions based on the phase, a simple decision tree classifier achieved 93.2% accuracy in distinguishing between the recompensated and decompensated phases, assessed using a LOSO evaluation.

Using LSTM techniques, an innovative predictive model framework for the diagnosis of heart failure is presented by Jin et al. [45]. When it comes to diagnosing heart failure, our algorithm performs better than well-known techniques like LR, RF, and AdaBoost. We describe patient diagnostic events in the experimental information' analysis and preprocessing by using word embedding vectors and one-hot encoding. Our analysis emphasizes how crucial it is to preserve clinical records' sequential structure. The improved performance of the LSTM model is demonstrated by the fact that its disease prediction accuracy outperforms that of the LR, RF, and AdaBoost algorithms.

### 2.4.4 ECG-based detection of cardiovascular ailments

There are numerous methods that utilize ECG for heart disease detection. ECG-based detection is a non-invasive method for identifying heart illness that measures cardiac health, heart rate, and heartbeat. Numerous studies indicate that the goal of HDPM can be achieved with better performance when the analysis is developed using models of ML. Recent developments in ML demonstrate the benefits of discriminative classifiers for autonomous identification of cardiac illness. According to earlier research, decision-making tools can effectively use different ML algorithms to forecast cardiac disease based on personally identifiable information.

Rajkumar and Reena used ML methods to forecast cardiac illness [46]. The tool named Tanagra is utilized for data classification, and a 10-fold CV is utilized to assess the data. S. Kodati and R. Vivekanandam [47] also analyzed various ML algorithms for classification. A set of 303 records from the UCI repository is used in this study. Based on the comparison result, it was observed that Decision Table outperforms other algorithms in terms of accuracy. According to Soganciogiu et al. [48], the TAR component of the NB classifier is suitable to diagnose cardiac disease. A TA is used to preprocess the data, and to find TARs, a temporal pattern mining technique is employed to find recurring temporal correlations between TAs. A survey of ML algorithms for estimating a person's risk of disease related to heart based on various features like age, gender, pulse rate, and cholesterol level, was carried out. The author can use techniques like neural networks, Naïve Bayes, and the K-nearest neighbor algorithm to improve the accuracy for risk assessments and trait analysis. Increasing accuracy can be achieved even with a limited number of features by using different methods [49].

Heart disease and PAC were using machine learning techniques to handle patient records. To analyze different machine learning techniques, map reduce programming is utilized and provided in a graphical format [50]. Moreover, it can be expanded to diagnose various illnesses including brain tumors, diabetes, cancer, and so forth.

## 2.5 Challenges and Opportunities

Medical diagnostic systems stand to benefit greatly from the advancement and discussion of AI algorithms, as demonstrated in this research. The diagnosis of AF is the most crucial and difficult step in its management. However, in order to fully utilize AI's potential for extracting new insights from the associated medical data and to develop a computationally efficient ML-based solution for automatic diagnosis of AF with accurate results, AI-based diagnostic systems need to address the following important issues.

### 2.5.1 Increased requirement for explainable AI

AI models are often criticized for their opaque decision-making processes. Explainable AI addresses this by clarifying and reasoning about the behavior of the AI methods used particularly deep learning. Additionally, to identify pattern recognition issues, AI systems should incorporate causal models of the world to make it more explainable and understandable. This is especially crucial for AI applications in medical diagnostics. Researchers emphasize the need to go beyond explainable AI, suggesting that causability will eventually lead to explainable diagnoses that include quality measurements for the explanations provided.

### 2.5.2 Detection at the initial stage or auto-detection

The best-case scenario is to identify the possible risk of AF before it manifests. In research, machine learning models are mainly used to identify AF at the onset or just before the occurrence. The energy constraints of continuous monitoring equipment and the dearth of effective ML-based models for AF prediction are two significant obstacles to early auto-detection. Such models are necessary to be able to handle low-energy monitoring devices in addition to having better accuracy in predicting the probability of AF far in advance. The creation of energy-efficient monitoring devices that can generate their own power can aid in overcoming energy-related constraints. The creation of effective ML models can greatly benefit from the data fusion of several sources. Big data analytics can assist in the fusion of data to identify significant biomarkers unique to a certain disease from multiple sources. Then, by using ML algorithms, these enriched data can be utilized to create models that anticipate AF well in advance.

### 2.5.3 Model training

The efficacy of ML algorithms is mostly contingent upon the accessibility of superior training models to attain the necessary diagnostic proficiency. Moreover, since huge datasets are essential to AI-based medical applications, the issue of data scarcity is crucial. It is yet unknown whether information augmentation and picture synthesis – two more techniques being used to deliver more annotated data – are appropriate for AI-based medical evaluations.

### 2.5.4 Requirement of alternate technology like wearable devices

ECG is frequently the primary diagnostic method for AF. However, the investigation of novel sensing technology avenues ought to be considered in situations when the development and implementation of increasingly complex ECG approaches are necessary. It would be beneficial to investigate technologies such as IEM, MRA, photoplethysmography, echocardiography, augmented reality, accelerometry, and radio frequency and microwave sensing systems. These technologies have been applied to similar healthcare issues. The other related data can also be collected by wearable devices. However, patients are wary and reluctant to adopt implants. Despite significant advancements in sensor technology, implant procedures can result in infections or even implant failures. The durability and quality of wearables also call for significant improvement. Wearable technology should be capable of performing well in a variety of conditions, including hot and muggy settings. This would provide constant and sufficient monitoring without compromising performance.

### 2.5.5 Lack of interoperability and standardization

Standardization is another essential component needed for an ML-based system for detecting and predicting AF, to be economically viable. Patients with AF may have different conditions. Furthermore, providers can use a collection of AI algorithms selected from a range of methods in a variety of ways to create a broad range of diagnostic products. However, they might not follow the standards and recommendations for suitable interfaces and associated protocols across different computer frameworks. This leads to concerns with interoperability. In order to manage system diversity, it will act swiftly to define the technological needs for AI-based diagnosis and medicine. Therefore, care should be taken while choosing biomarkers, gathering information, extracting features, and putting machine learning algorithms into practice so that they function well across a wide range of population types. It is a difficult effort to develop a solution of this kind that is extremely accurate, computationally cheap, and suitable for a wide range of users. In this regard, a number of technical and medical bodies, including the World Health Professionals Alliance, the World Health Organization, and the International Organization for Standardization's AI group, can work together.

### 2.5.6 Security

Security has two aspects: first, the security of the data of the patient in the creation of auto-diagnosis, and secondly, the secure diagnosis of AF. When data is saved in a shared DB and is sent over the Internet, it becomes even more crucial. It is crucial to protect each participant's personal information when using a shared DB to gain access to additional information that can be used to create smarter solutions. However, the exploitation of these approaches can also reveal personal information or the participant's identity. For the safe transfer and storage of data, more sophisticated data encryption algorithms are therefore required. However, one limitation associated with encryption is that even basic mathematical operations yield rather imprecise results for encrypted data.

In general, AI and, in particular, DL techniques are very tailored to the application; that is, a model created for one diagnosis might not be appropriate for another. The algorithms usually need to be trained again using the relevant medical data in order to be used for different diseases; otherwise, an inaccurate diagnosis is unavoidable. Additionally, incorrect hyperparameter selections might cause a significant shift in the model's performance, leading to a flawed diagnosis, even with a small adjustment. Another essential element of safe diagnostics is the requirement to protect the diagnostic systems from malevolent actors. The attackers break into the system by taking advantage of the AI algorithms' features.

## 2.6 Conclusion

An important development in the battle against cardiovascular disorders, such as AF, is the use of AI and ML in medical diagnosis. The enormous potential of AI-based systems to improve early detection, increase diagnostic precision, and lower healthcare costs has been emphasized in this article, tackling some of the most urgent issues facing the industry. But in order to fully reap these benefits, many obstacles must be overcome, including the need for explainable AI, the creation of accurate and energy-efficient prediction models, the accessibility of high-quality training data, and the development of wearable technology. The scalability and dependability of AI-driven systems also depend on strong data protection measures, platform interoperability, and standards. In order to overcome these obstacles, promote innovation, and guarantee fair access to AI-enabled diagnostics, cooperation across technological, medical, and regulatory organizations will be essential. AI and ML have the ability to transform the diagnosis and treatment of AF by addressing these obstacles, which would eventually lead to better patient outcomes and a notable decrease in the worldwide burden of heart diseases. Unlocking these technologies' transformational potential will need ongoing study and multidisciplinary cooperation.

## References

[1] URL: https://www.who.int/health-topics/cardiovascular-diseases#tab=tab_1

[2] M. M. Ahsan and Z. Siddique, "Machine learning-based heart disease diagnosis: A systematic literature review," *Artificial Intelligence in Medicine,* vol. 128, p. 102289, 2022.

[3] J. Nahar, T. Imam, K. S. Tickle and Y.-P. P. Chen, "Computational intelligence for heart disease diagnosis: A medical knowledge driven approach," *Expert systems with applications,* vol. 40, p. 96–104, 2013.

[4] S. V. Sree, D. N. Ghista and K.-H. Ng, "Cardiac arrhythmia diagnosis by HRV signal processing using principal component analysis," *Journal of Mechanics in Medicine and Biology,* vol. 12, p. 1240032, 2012.

[5] N. Liu, Z. Lin, J. Cao, Z. Koh, T. Zhang, G.-B. Huang, W. Ser and M. E. H. Ong, "An intelligent scoring system and its application to cardiac

arrest prediction," *IEEE Transactions on Information Technology in Biomedicine,* vol. 16, p. 1324–1331, 2012.

[6] P. Khan, M. F. Kader, S. R. Islam, A. B. Rahman, M. S. Kamal, M. U. Toha and K.-S. Kwak, "Machine learning and deep learning approaches for brain disease diagnosis: principles and recent advances," *Ieee Access,* vol. 9, p. 37622–37655, 2021.

[7] R. Indrakumari, T. Poongodi and S. R. Jena, "Heart disease prediction using exploratory data analysis," *Procedia Computer Science,* vol. 173, p. 130–139, 2020.

[8] S. Nashif, M. R. Raihan, M. R. Islam and M. H. Imam, "Heart disease detection by using machine learning algorithms and a real-time cardiovascular health monitoring system," *World Journal of Engineering and Technology,* vol. 6, p. 854–873, 2018.

[9] M. Kelly, R. Longjohn and K. Nottingham, "The UCI machine learning repository," *URL https://archive. ics. uci. edu,* 2023.

[10] M. Zubaid, W. A. Rashed, A. A. Alsheikh-Ali, W. AlMahmeed, A. Shehab, K. Sulaiman, I. Al-Zakwani, A. AlQudaimi, N. Asaad and H. Amin, "Gulf survey of atrial fibrillation events (Gulf SAFE) design and baseline characteristics of patients with atrial fibrillation in the arab middle East," *Circulation: Cardiovascular Quality and Outcomes,* vol. 4, p. 477–482, 2011.

[11] G. Moody, A. Goldberger, S. McClennen and S. Swiryn, "Predicting the onset of paroxysmal atrial fibrillation: The Computers in Cardiology Challenge 2001," in *Computers in Cardiology 2001. Vol. 28 (Cat. No. 01CH37287)*, 2001.

[12] G. B. Moody and R. G. Mark, "The impact of the MIT-BIH arrhythmia DB," *IEEE engineering in medicine and biology magazine,* vol. 20, p. 45–50, 2001.

[13] A. L. Goldberger, L. A. N. Amaral, L. Glass, J. M. Hausdorff, P. C. Ivanov, R. G. Mark, J. E. Mietus, G. B. Moody, C.-K. Peng and H. E. Stanley, "PhysioBank, PhysioToolkit, and PhysioNet: components of a new research resource for complex physiologic signals," *circulation,* vol. 101, p. e215–e220, 2000.

[14] G. E. Moody, "Spontaneous termination of atrial fibrillation: a challenge from physionet and computers in cardiology 2004," in *Computers in Cardiology, 2004*, 2004.

[15] G. D. Clifford, C. Liu, B. Moody, H. L. Li-wei, I. Silva, Q. Li, A. E. Johnson and R. G. Mark, "AF classification from a short single lead ECG recording: The PhysioNet/computing in cardiology challenge 2017," in *2017 Computing in Cardiology (CinC)*, 2017.

[16] J.-Y. Le Heuzey, G. Breithardt, J. Camm, H. Crijns, P. Dorian, P. R. Kowey, I. Merioua, E. N. Prystowsky, P. J. Schwartz, C. Torp-Pedersen and others, "The RecordAF study: design, baseline data, and profile of patients according to chosen treatment strategy for atrial fibrillation," *The American journal of cardiology,* vol. 105, p. 687–693, 2010.

[17] R. K. Halder, URL: https://ieee-dataport.org/documents/cardio vascular-disease-dataset, 2022.

[18] S. Liu, J. Shao, T. Kong and R. Malekian, "ECG arrhythmia classification using high order spectrum and 2D graph fourier transform," *Applied Sciences,* vol. 10, p. 4741, 2020.

[19] L. B. Marinho, N. de MM Nascimento, J. W. M. Souza, M. V. Gurgel, P. P. Rebouças Filho and V. H. C. de Albuquerque, "A novel electrocardiogram feature extraction approach for cardiac arrhythmia classification," *Future Generation Computer Systems,* vol. 97, p. 564–577, 2019.

[20] C. Zou, A. Müller, U. Wolfgang, D. Rückert, P. Müller, M. Becker, A. Steger and E. Martens, "Heartbeat classification by random forest with a novel context feature: a segment label," *IEEE Journal of Translational Engineering in Health and Medicine,* vol. 10, p. 1–8, 2022.

[21] W. Ji and D. Zhu, "ECG classification exercise health analysis algorithm based on GRU and convolutional neural network," *IEEE Access,* 2024.

[22] F. Khan, X. Yu, Z. Yuan and A. U. Rehman, "ECG classification using 1-D convolutional deep residual neural network," *Plos one,* vol. 18, p. e0284791, 2023.

[23] Y. Lu, M. Jiang, L. Wei, J. Zhang, Z. Wang, B. Wei and L. Xia, "Automated arrhythmia classification using depthwise separable convolutional neural network with focal loss," *Biomedical Signal Processing and Control,* vol. 69, p. 102843, 2021.

[24] E. Merdjanovska and A. Rashkovska, "Cross-DB generalization of deep learning models for arrhythmia classification," in *2021 44th International Convention on Information, Communication and Electronic Technology (MIPRO)*, 2021.

[25] P. G. Gaddam, R. V. Sreehari and others, "Automatic classification of cardiac arrhythmias based on ECG signals using transferred deep learning convolution neural network," in *Journal of Physics: Conference Series*, 2021.

[26] D. K. Atal and M. Singh, "Arrhythmia classification with ECG signals based on the optimization-enabled deep convolutional neural network," *Computer Methods and Programs in Biomedicine,* vol. 196, p. 105607, 2020.

[27] Z. Zheng, Z. Chen, F. Hu, J. Zhu, Q. Tang and Y. Liang, "An automatic diagnosis of arrhythmias using a combination of CNN and LSTM technology," *Electronics,* vol. 9, p. 121, 2020.

[28] S. Nurmaini, M. N. Rachmatullah, A. I. Sapitri, A. Darmawahyuni, A. Jovandy, F. Firdaus, B. Tutuko and R. Passarella, "Accurate detection of septal defects with fetal ultrasonography images using deep learning-based multiclass instance segmentation," *IEEE Access,* vol. 8, p. 196160–196174, 2020.

[29] L. Xu, M. Liu, J. Zhang and Y. He, "Convolutional-neural-network-based approach for segmentation of apical four-chamber view from fetal echocardiography," *IEEE Access,* vol. 8, p. 80437–80446, 2020.

[30] T. Rahman, M. K. A. A. Al-Ruweidi, M. S. I. Sumon, R. Y. Kamal, M. E. H. Chowdhury and H. C. Yalcin, "Deep Learning Technique for Congenital Heart Disease Detection using Stacking-based CNN-LSTM Models from Fetal Echocardiogram: A Pilot Study," *IEEE Access,* 2023.

[31] S. Ziani, Y. Farhaoui and M. Moutaib, "Extraction of fetal electrocardiogram by combining deep learning and SVD-ICA-NMF methods," *Big Data Mining and Analytics,* vol. 6, p. 301–310, 2023.

[32] Y. Gong, Y. Zhang, H. Zhu, J. Lv, Q. Cheng, H. Zhang, Y. He and S. Wang, "Fetal congenital heart disease echocardiogram screening based on DGACNN: adversarial one-class classification combined with video transfer learning," *IEEE transactions on medical imaging,* vol. 39, p. 1206–1222, 2019.

[33] S. Qiao, S. Pang, G. Luo, S. Pan, T. Chen and Z. Lv, "FLDS: An intelligent feature learning detection system for visualizing medical images supporting fetal four-chamber views," *IEEE Journal of Biomedical and Health Informatics,* vol. 26, p. 4814–4825, 2021.

[34] B. Xiao, Y. Xu, X. Bi, W. Li, Z. Ma, J. Zhang and X. Ma, "Follow the sound of children's heart: a deep-learning-based computer-aided pediatric CHDs diagnosis system," *IEEE Internet of Things Journal,* vol. 7, p. 1994–2004, 2019.

[35] M. Alkhodari, L. J. Hadjileontiadis and A. H. Khandoker, "Identification of congenital valvular murmurs in young patients using deep learning-based attention transformers and phonocardiograms," *IEEE Journal of Biomedical and Health Informatics, vol. 28 (4),* p. 1803-1814, 2024.

[36] A. Ungureanu, A.-S. Marcu, C. L. Patru, D. Ruican, R. Nagy, R. Stoean, C. Stoean and D. G. Iliescu, "Learning deep architectures for the interpretation of first-trimester fetal echocardiography (LIFE)-a study protocol for developing an automated intelligent decision support system for

early fetal echocardiography," *BMC Pregnancy and Childbirth,* vol. 23, p. 20, 2023.

[37] W. Ning, S. Li, D. Wei, L. Z. Guo and H. Chen, "Automatic detection of congestive heart failure based on a hybrid deep learning algorithm in the internet of medical things," *IEEE Internet of Things Journal,* vol. 8, p. 12550–12558, 2020.

[38] A. M. Qadri, A. Raza, K. Munir and M. S. Almutairi, "Effective feature engineering technique for heart disease prediction with machine learning," *IEEE Access,* vol. 11, p. 56214–56224, 2023.

[39] L. Zou, Z. Huang, X. Yu, J. Zheng, A. Liu and M. Lei, "Automatic detection of congestive heart failure based on multiscale residual unet++: From centralized learning to federated learning," *IEEE Transactions on Instrumentation and Measurement,* vol. 72, p. 1–13, 2022.

[40] E. Prabhakararao and S. Dandapat, "Congestive heart failure detection from ECG signals using deep residual neural network," *IEEE Transactions on Systems, Man, and Cybernetics: Systems,* vol. 53, p. 3008–3018, 2022.

[41] L. Wang, W. Zhou, Q. Chang, J. Chen and X. Zhou, "Deep ensemble detection of congestive heart failure using short-term RR intervals," *IEEE Access,* vol. 7, p. 69559–69574, 2019.

[42] L. Zhao, C. Liu, S. Wei, C. Liu and J. Li, "Enhancing detection accuracy for clinical heart failure utilizing pulse transit time variability and machine learning," *IEEE Access,* vol. 7, p. 17716–17724, 2019.

[43] A. Ishaq, S. Sadiq, M. Umer, S. Ullah, S. Mirjalili, V. Rupapara and M. Nappi, "Improving the prediction of heart failure patients' survival using SMOTE and effective data mining techniques," *IEEE access,* vol. 9, p. 39707–39716, 2021.

[44] M. Gjoreski, A. Gradišek, B. Budna, M. Gams and G. Poglajen, "Machine learning and end-to-end deep learning for the detection of chronic heart failure from heart sounds," *Ieee Access,* vol. 8, p. 20313–20324, 2020.

[45] B. Jin, C. Che, Z. Liu, S. Zhang, X. Yin and X. Wei, "Predicting the risk of heart failure with EHR sequential data modeling," *Ieee Access,* vol. 6, p. 9256–9261, 2018.

[46] A. Rajkumar and G. S. Reena, "Diagnosis of heart disease using datamining algorithm," *Global journal of computer science and technology,* vol. 10, p. 38–43, 2010.

[47] S. Kodati and D. R. Vivekanandam, "A comparative study on open source data mining tool for heart disease," *International Journal of Innovations & Advancement in Computer Science,* vol. 7, p. 81–87, 2018.

[48] E. Sogancioglu, K. Murphy, E. Calli, E. T. Scholten, S. Schalekamp and B. Van Ginneken, “Cardiomegaly detection on chest radiographs: Segmentation versus classification,” *IEEE Access,* vol. 8, p. 94631–94642, 2020.

[49] F. Z. Abdeldjouad, M. Brahami and N. Matta, “A hybrid approach for heart disease diagnosis and prediction using machine learning techniques,” in *The Impact of Digital Technologies on Public Health in Developed and Developing Countries: 18th International Conference, ICOST 2020, Hammamet, Tunisia, June 24–26, 2020, LNCS Proceedings, vol. 12157, Springer*, 2020.

[50] A. Tyagi and R. Mehra, “Intellectual heartbeats classification model for diagnosis of heart disease from ECG signal using hybrid convolutional neural network with GOA,” *SN Applied Sciences,* vol. 3, p. 265, 2021.

# 3

# Stacked Machine Learning for Accurate Prediction of Dual-Band Circular Patch Antenna Resonant Frequencies

**Mitesh Upreti[1] and Sanjay Mathur[2]**

[1]Department of Electronics and Communication Engineering, GBPUA&T, India
[2]Department of Electronics and Communication Engineering, GBPUA&T, India

**Email:** 51154@gbpuat.ac.in; sanjaymathur.ece@gbpuat-tech.ac.in

## Abstract

This chapter presents a machine learning-based approach for accurately predicting the resonant frequencies of dual-band circular patch antennas. A dataset was generated using high-frequency structure simulator simulations within the 2–7 gigahertz frequency range, considering geometric parameters such as cut width, feed width, and side length, along with the number of bands as input features. The outputs comprise two resonant frequencies corresponding to dual-band operation. Various regression models – random forest, extreme gradient boosting, and a neural network – were developed and evaluated. To enhance prediction accuracy, ensemble strategies including model blending and model stacking were implemented. Among the individual models, Random Forest yielded the best performance (root mean square error = 0.2106, coefficient of determination = 0.9837), while the stacked ensemble outperformed all others (root mean square error = 0.1936, coefficient of determination = 0.9860). The results demonstrate that stacked learning significantly improves frequency prediction accuracy.



DOI: 10.1201/9788743812739-3

## 3.1 Introduction

Antennas are fundamental to modern wireless communication, facilitating efficient transmission and reception of signals [1]. Among different types, microstrip patch antennas are widely adopted due to their compact size, lightweight construction, and ease of fabrication. These antennas are particularly useful in applications such as 5G networks, satellite communication, radar systems, and IoT devices [2,3]. However, designing a dual-band antenna with optimal performance requires precise tuning of multiple parameters, making the design process complex and time-consuming.

To enhance the efficiency of antenna design, machine learning (ML) and artificial intelligence (AI) have emerged as promising alternatives to traditional computational techniques like the finite element method (FEM) and method of moments (MoM) [4,5]. While conventional numerical methods demand extensive computational resources, ML-based models can quickly analyze large datasets and provide accurate predictions, thereby streamlining the optimization process [6,7,8].

In this research, dual-band resonant frequency prediction for circular patch antennas [9] is explored using ensemble learning techniques [10]. The antenna under study is modeled using HFSS (high-frequency structure simulator) with a 60*60 mm dimension and operates in the 2 to 7 GHz frequency range. After performing multiple simulations and optimizations, a dataset is generated, including design parameters such as cut width, feed width, side length, and resonant frequencies.

A crucial aspect of this study is identifying the number of operating bands (no. of bands), which is determined using a Python-based approach that evaluates resonant frequency 1 and 2. The computed no. of bands is then incorporated as an input feature in the machine learning models. To improve prediction accuracy, ensemble learning techniques like random forest [11], XGBoost, and neural networks (NNs) [12] are employed. Additionally, blending and stacking strategies are applied to enhance model performance further [13,14].

The effectiveness of the models is assessed using root mean square error (RMSE) and $R$ score, where the stacked model achieves superior accuracy (RMSE = 0.2004, $R$ = 0.9854), surpassing individual models and blended techniques. This study demonstrates the potential of AI-driven methodologies in antenna design, enabling faster and more precise predictions, which are essential for optimizing performance in advanced wireless communication systems.

The following sections of this chapter are organized as follows: Section 3.2 reviews related work, Section 3.3 details the antenna design and simulation process, Section 3.4 presents the machine learning approach using the circular patch antenna, Section 3.5 discusses the results and performance evaluation, and Section 3.6 concludes the study while suggesting its limitations.

## 3.2 Related Work and Contribution Significance

Machine learning (ML) techniques have been increasingly applied to antenna design and optimization. Elmisilmani et al. [15] provided a comprehensive review of ML applications in antenna design, highlighting the potential of regression and classification models. Al-Gburi et al. [16] further discussed advancements in ML techniques for antenna design, emphasizing data-driven modeling and optimization algorithms.

Han et al. [17] employed deep kernel learning for resonant-frequency modeling of microstrip antennas, demonstrating improved prediction accuracy. Ong et al. [18] utilized support vector machines (SVM) to classify circular patch antennas based on resonance and impedance matching, achieving a 96% accuracy rate. Varma and Ghosh [19] proposed a hybrid model combining neural networks and SVM for dual-band proximity-coupled microstrip antennas, offering flexibility in design across various frequency bands.

Chen and Tian [20] developed a CUDA-based artificial neural network trained by particle swarm optimization for modeling the resonant frequency of rectangular microstrip antennas, achieving significant speedup in training time. Pachori et al. [21] applied deep neural networks, random forest, and XGBoost to predict the performance of dielectric resonator-based MIMO antennas and validated their models with experimental results.

Sutrakar et al. [22] utilized regression-based ML models for designing cavity-backed slotted antennas, enabling rapid prediction of optimal antenna configurations over a wide frequency band. Nuray and Ak¸cam [23] designed a dual-band F-shaped RFID antenna using various ML algorithms, with random forest achieving the highest prediction accuracy.

Despite these advancements, most prior works focus on single-band antennas or classification tasks. In contrast, our study uniquely formulates the dual-band circular patch antenna design problem as a multi-output regression task. We apply advanced ensemble techniques – blending and stacking – which aggregate multiple base learners (random forest, XGBoost, neural networks) to enhance prediction accuracy. Our approach significantly outperforms individual models, as evidenced by lower RMSE and higher $R^2$ scores, providing a scalable and precise framework for data-driven antenna modeling.

## 3.3 Antenna Design

The proposed antenna is designed on an FR-4 epoxy substrate, which has a relative permittivity ($\varepsilon_r$) of 4.4, a dielectric loss tangent of 0.02, and a thickness of 1.6 mm. The overall dimensions are 60 mm × 60 mm, featuring a circular patch topology. The key design parameters used in the dataset are extracted from ANSYS HFSS, a widely used 3D electromagnetic simulation tool [24], and are as follows:

**Radius (mm):** Refers to the radius of the circular patch, directly influencing the fundamental resonant frequency due to its impact on surface current distribution.

**Cut Width (mm):** Denotes the width of the slot introduced in the patch near the feed for impedance matching.

**Feed Width (mm):** The width of the microstrip feed line that excites the patch.

**Number of Bands:** Indicates whether the antenna supports single-band or dual-band operation.

**Resonant Frequencies (GHz):** These are output parameters from HFSS simulations representing the fundamental and secondary resonant frequencies of the antenna structure.

The resonant frequencies derived from simulations span a wide range (2–7 GHz), adding variability to the dataset and thereby influencing the regression performance of ML algorithms [4, 25].

Figure 3.1 presents the structure of the circular patch antenna with labeled input parameters. These inputs were used to generate the simulation dataset, which serves as the foundation for the data-driven modeling approach discussed in subsequent sections [26,27].

## 3.4 Machine Learning using Circular Patch Antenna

Machine learning plays a crucial role in predicting the resonant frequencies of circular patch antennas by analyzing the relationship between design parameters and frequency response [15, 28] as represented by the block diagram in Figure 3.2. This section outlines the data preparation, algorithms used, and the implementation of ensemble learning techniques to enhance prediction accuracy.

The dataset was generated through HFSS simulations, optimizing key antenna parameters such as cut width, feed width, and radius. These features,

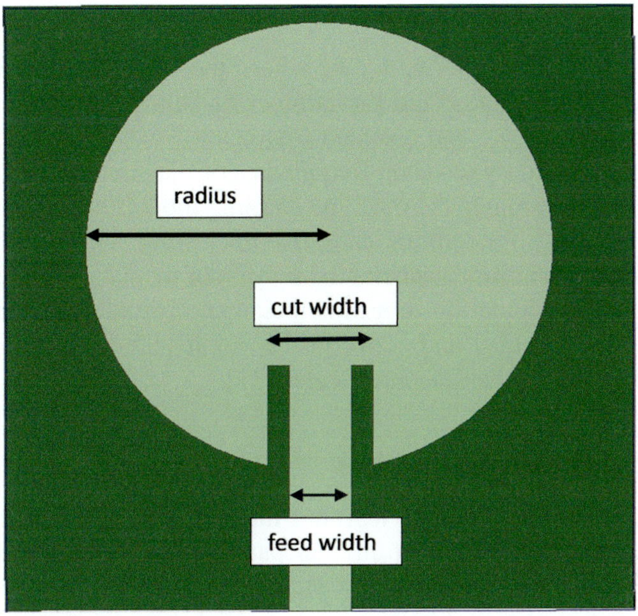


**Figure 3.1** Geometry of the dual-band circular patch antenna with labeled input features.

along with the number of bands, were used as input variables. Python code was used to classify antennas into single-band or dual-band categories using resonant frequencies, while regression models, including random forest [11], XGBoost [29], and neural networks [12], were used to predict resonant frequencies. Additionally, ensemble learning techniques – blending and stacking – were implemented to enhance model performance [10, 30].

The following subsections describe the data preparation process, the machine learning algorithms used, and the ensemble learning approach in greater detail.

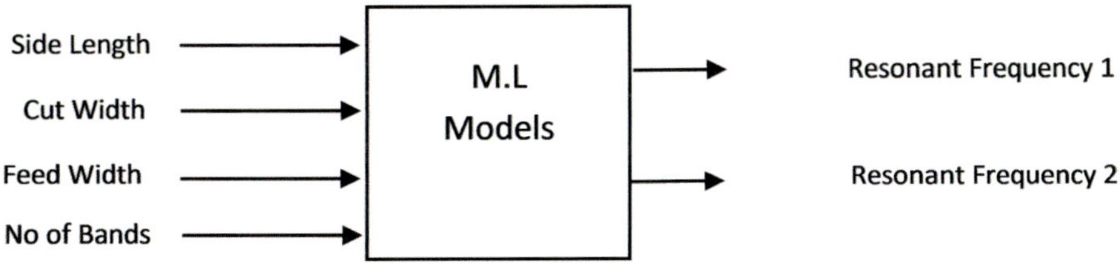


**Figure 3.2** Machine learning model for the prediction of resonant frequencies.

### 3.4.1 Training and testing dataset

The dataset consists of 447 samples, where the input features include cut length, feed length, radius, and the number of bands. The output variables are resonant frequency 1 and resonant frequency 2, which represent the fundamental and secondary resonant frequencies of the circular patch antenna.

To train and evaluate the machine learning models effectively, the dataset is split into 80% for training and 20% for testing. This division ensures that the models learn from a substantial portion of the data while maintaining a separate set for validation to assess their generalization performance. To improve model accuracy and convergence, input features and target values are normalized using the Standard-Scaler [31].

### 3.4.2 Machine learning models

In this study, various machine learning models were implemented to predict the resonant frequencies of circular patch antennas based on geometric parameters. The models considered include random forest, XGBoost, and artificial neural networks (ANNs). Each model was optimized to improve predictive performance by adjusting hyperparameters and preprocessing techniques [32].

#### 3.4.2.1 Random forest regressor

Random forest (RF) [11] is an ensemble learning method that constructs multiple decision trees and aggregates their outputs to enhance prediction accuracy. It is well-suited for capturing non-linear relationships and reducing overfitting through bootstrap aggregation.

The RF model was trained using 300 decision trees (estimators) with a maximum depth of 15. These hyperparameters were optimized to balance complexity and generalization.

$$RF = \frac{1}{N}\sum_{i=0}^{i=N} T_i\left(X\right) \tag{3.1}$$

where $T_i(X)$ represents the prediction of the $i$th decision tree, and $N$ is the total number of trees.

#### 3.4.2.2 XGBoost regressor

Extreme gradient boosting (XGBoost) [29] is a powerful boosting algorithm that enhances weak learners by minimizing errors iteratively. For this study, the XGBoost model was configured with 500 estimators, a learning rate of 0.02, and a maximum depth of 10.

$$y_{pred} = \sum_{i=1}^{i=N} \eta f_i(X) \tag{3.2}$$

where $f_i(X)$ represents the output of the $i$th weak learner (decision tree). $\eta$ is the learning rate, controlling how much each tree contributes to the final prediction. T is the total number of trees in the model

#### 3.4.2.3 Artificial neural network (ANN)

Artificial neural networks (ANNs) [33] are powerful deep learning architectures capable of modeling complex, non-linear relationships within data. In this work, a feedforward ANN was implemented to predict the dual-band resonant frequencies of circular patch antennas. The model leverages multiple hidden layers and employs ReLU (rectified linear unit) activation functions for non-linearity, along with dropout for regularization and overfitting prevention.

The mathematical formulation of the network's output is given by:

$$y_{pred} = W_2 \cdot ReLU\left(W_1 X + b_1\right) + b_2 \tag{3.3}$$

where:

$X$ is the input feature vector,
$W_1$, $W_2$ are the weight matrices for the first and second layers,
$b_1$, $b_2$ are the corresponding bias terms,
$ReLU(z) = \max(0, z)$ is the activation function applied element-wise.

The neural network architecture is defined as follows:

**Input Layer:** 4 neurons corresponding to the input features

**Hidden Layers:**

- First layer: 128 neurons, ReLU activation, Dropout (rate = 0.2),
- Second layer: 64 neurons, ReLU activation.
- Third layer: 32 neurons, ReLU activation

**Output Layer:** 2 neurons to simultaneously predict the two resonant frequencies

**Optimizer:** Adam optimizer with a learning rate of 0.01

**Loss Function:** Mean squared error (MSE)

**Training Parameters:** 300 epochs with a batch size of 8

This architecture was selected based on its ability to balance complexity and generalization, enabling accurate predictions without excessive overfitting. Similar approaches have been successfully used in antenna modeling tasks [34, 17], further validating the suitability of deep learning techniques for electromagnetic prediction problems.

### 3.4.3 Ensembling techniques in machine learning

Ensemble learning combines multiple models to enhance prediction performance and reduce errors [35]. This study utilized blending and stacking methods.

#### 3.4.3.1 Blending approach

Blending is an ensemble technique that combines the predictions of multiple base models through a weighted average to improve overall accuracy and robustness [36]. Unlike stacking, which employs a meta-learner to learn optimal combinations, blending utilizes weights that are predefined or optimized using a validation-based strategy.

The final prediction $y_{\text{pred}}$ is computed as:

$$y_{pred} = w_1 \cdot Y_{RF} + w_2 \cdot Y_{NN} + w_3 \cdot Y_{XGB} \tag{3.4}$$

where:

$Y_{RF}$, $Y_{NN}$, $Y_{XGB}$ are predictions from the random forest, neural network, and XGBoost models, respectively.

$w_1$, $w_2$, $w_3$ are the blending weights assigned to each model's prediction.

##### 3.4.3.1.1 Weight calculation method

The optimal weights were determined through grid search optimization:

1. Candidate weight combinations ($w_1$, $w_2$, $w_3$) were generated under the constraint $w_1 + w_2 + w_3 + 1$.
2. For each combination, the blended predictions were computed on a validation dataset.
3. The Root Mean Squared Error (RMSE) was evaluated for each blend.
4. The combination yielding the lowest validation MSE was selected as the optimal weight configuration.

This approach allows the model to exploit the individual strengths of base learners while reducing generalization error in the final predictions.

#### 3.4.3.2 Stacking approach

Stacking [37] is an ensemble method where predictions from multiple base models are combined and passed to a meta-learner for final prediction. This technique enhances accuracy by leveraging the strengths of each base model.

The combined feature vector is:

$$S = [Y_{RF}, Y_{NN}, Y_{XGB}] \tag{3.5}$$

and the stacked prediction is:

$$Y_{stack} = f(S) = Linear\ Regression \tag{3.6}$$

Linear regression was used as the meta-learner to minimize residual error. Stacking has shown strong performance in recent regression studies [38, 39, 40], aligning with our findings in resonant frequency prediction.

### 3.4.4 Sample dataset

The dataset used for training the machine learning models consists of optimized geometric parameters of circular patch antennas, including cut width, feed width, radius, and number of bands. These parameters were used as input features, while the corresponding resonant frequencies (resonant freq 1 and resonant freq 2) served as output variables.

The data presented in Table 3.1 were generated using parametric simulations in *Ansys HFSS*, where each antenna design was evaluated over the 2–7 GHz frequency range. The number of bands was determined using a custom *Python* script, which identified the presence of one or two resonant frequency peaks based on thresholding criteria in the simulated S-parameter plots.

**Table 3.1** Sample dataset generated using HFSS.

| Radius (mm) | Cut Width (mm) | Feed Width (mm) | No. of Bands | Resonant Freq 1 (GHz) | Resonant Freq 2 (GHz) |
|---|---|---|---|---|---|
| 14.0 | 5 | 2 | 1 | 5.076 | 0.000 |
| 17.0 | 5 | 2 | 2 | 4.110 | 5.691 |
| 14.5 | 6 | 2 | 2 | 4.940 | 6.907 |
| 11.5 | 7 | 3 | 1 | 6.170 | 0.000 |
| 23.0 | 8 | 4 | 2 | 3.130 | 4.353 |

## 3.5 Result and Discussion

In this section, we evaluate the performance of the proposed machine learning models in predicting the dual-band resonant frequencies of circular patch antennas.

The predictive performance of the models was assessed using root mean square error (RMSE) and coefficient of determination ($R^2$) metrics. A comparative analysis was conducted between individual models – random forest, XGBoost, and neural network and ensemble techniques, including blending and stacking, to determine the most effective approach for accurate frequency prediction.

### 3.5.1 Performance of individual models

The performance of the random forest, XGBoost, and neural network models was first analyzed independently. Random forest provided stable predictions with relatively lower RMSE, while XGBoost performed well due to its boosting mechanism. Its error rate remained higher compared to the neural network. The neural network, though effective, had a slightly higher RMSE due to its dependency on hyperparameter tuning and training iterations.

### 3.5.2 Effectiveness of blending and stacking

To further enhance prediction accuracy and robustness, ensemble learning techniques – specifically blending and stacking were employed in this study.

Blending involves taking a weighted average of the predictions from multiple base models. In this case, predictions from random forest, XGBoost, and neural network were combined. The optimal weights were determined through a grid search-based optimization that minimized the root mean squared error (RMSE) on a validation set. The final selected weights were:

$w_1 = 0.55$ for random forest
$w_2 = 0.25$ for neural network
$w_3 = 0.20$ for XGBoost

This blending strategy led to improved performance compared to individual models, as it leveraged the complementary strengths of each learner while mitigating their individual weaknesses.

In stacking, the predictions of the base models (random forest, XGBoost, neural network) were used as input features for a second-level model – a linear regression meta-learner. This approach captures inter-model relationships

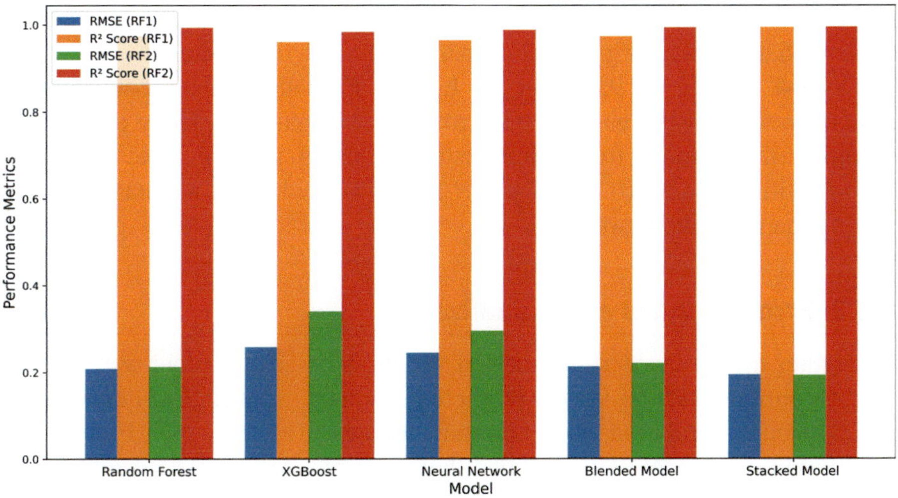


**Figure 3.3** Comparison of model performance for each resonant frequency.

and reduces both bias and variance. Among all tested approaches, stacking achieved the best results, with the lowest root mean squared error (RMSE) of 0.1936 and the highest coefficient of determination $R^2 = 0.9860$, clearly outperforming the base learners and the blended ensemble.

Figure 3.3 illustrates the performance comparison of different models for resonant frequency 1 and resonant frequency 2. Figure 3.4 presents the

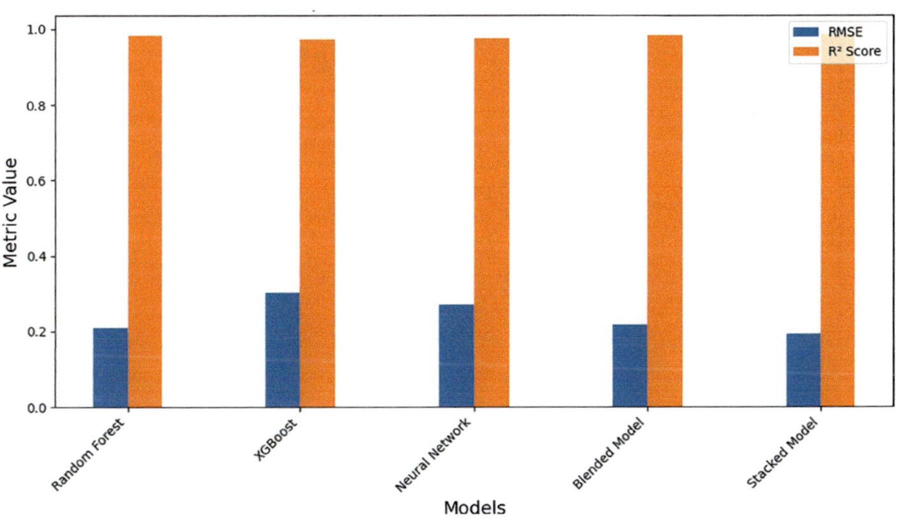


**Figure 3.4** Comparison of combined model performance.

overall performance of each model across both resonant frequencies. It is evident that the stacking model outperformed individual models, achieving the lowest RMSE (0.1936) and the highest *R* (0.9860). This confirms the effectiveness of using ensemble learning for resonant frequency prediction. Additionally, random forest demonstrated strong performance, achieving a competitive RMSE (0.2106) and *R* (0.9837), making it a reliable choice among the individual models.

### 3.5.3 Error analysis and visualization

To further assess the performance of the models, scatter plots were generated to compare actual versus predicted resonant frequencies, demonstrating a strong correlation with minimal deviation, as shown in Figure 3.5. Additionally, residual plots revealed that the errors followed a normal distribution, confirming the robustness and reliability of the models in predicting resonant frequencies (Figure 3.6). The tight clustering around the diagonal

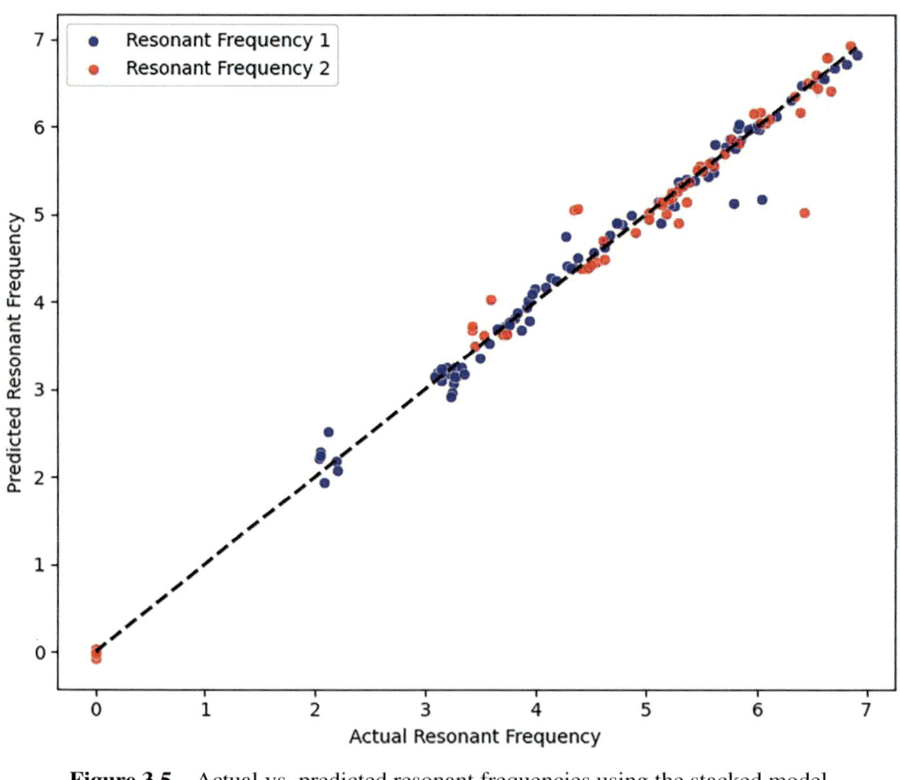


**Figure 3.5** Actual vs. predicted resonant frequencies using the stacked model.

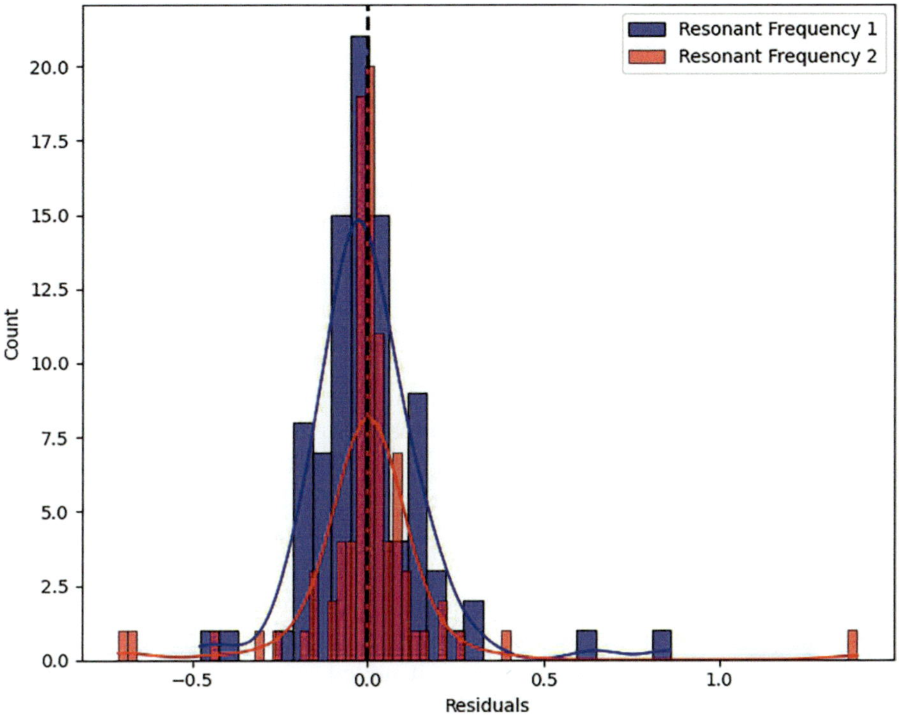


**Figure 3.6** Residual distribution plot for resonant frequencies 1 and 2.

line in the scatter plots highlights the accuracy of the predictions. Moreover, the residuals showed no evident pattern, suggesting that the models effectively captured the underlying data distribution without overfitting.

Table 3.2 showcases the predicted values obtained using the stacking model against the actual resonant frequencies. The small deviations indicate that the model successfully captures the underlying patterns in the dataset.

**Table 3.2** Comparison of actual vs. predicted resonant frequencies (GHz).

| Sample | Resonant Frequency 1 Actual | Resonant Frequency 1 Predicted | Resonant Frequency 2 Actual | Resonant Frequency 2 Predicted |
|---|---|---|---|---|
| **1** | 3.45 | 3.42 | 5.78 | 5.75 |
| **2** | 2.89 | 2.92 | 6.01 | 5.97 |
| **3** | 4.23 | 4.19 | 5.56 | 5.52 |
| **4** | 3.67 | 3.64 | 6.45 | 6.41 |
| **5** | 4.89 | 4.85 | 7.12 | 7.08 |

The comparative analysis of machine learning models demonstrated that ensemble techniques, particularly stacking, significantly improved the prediction accuracy of dual-band resonant frequencies in circular patch antennas. Among the individual models, Random Forest exhibited strong performance, with competitive RMSE and $R$ values. However, the stacking model outperformed all other approaches, achieving the lowest RMSE (0.1936) and highest $R$ (0.9860), highlighting the effectiveness of leveraging multiple predictive models.

## 3.6 Conclusion

This study applied ensemble learning techniques to predict the dual-band resonant frequencies of circular patch antennas. Random forest, XGBoost, and neural networks were employed as base models, with stacking achieving the best performance (RMSE = 0.1936, RRR = 0.9860). Predicted vs. actual plots and residual analysis confirmed the robustness of the models, underscoring the potential of ensemble methods for antenna design optimization. However, the dataset was relatively small and limited to specific geometric parameters, which may restrict generalization to other configurations. Moreover, while ensemble learning improved accuracy, it added computational complexity. Future work should expand the dataset and investigate lightweight ensemble strategies to balance accuracy with efficiency.

### Conflict of Interest

The authors confirm that there is no conflict of interest to declare for this publication.

### Acknowledgments

This research did not receive any specific grant from funding agencies in the public, commercial, or not-for-profit sectors. We would like to express their gratitude to the editor and anonymous reviewers for their valuable comments and suggestions, which significantly improved the quality of this work.

## References

[1] C. A. Balanis, *Antenna theory: Analysis and design*, 4th ed. John Wiley & Sons, 2015.

[2] J. J. Hall and P. S. Hall, *Handbook of microstrip antennas*. Peter Peregrinus, 1989.

[3] W. Li, J. Zhang, and X. Huang, "Dual-band microstrip antenna design using hybrid optimization algorithm for 5g applications," *IEEE Access*, vol. 9, pp. 102 345–102 356, 2021.

[4] H. El Misilmani, T. Naous, and S. Al Khatib, "A review on the design and optimization of antennas using machine learning algorithms and techniques," *International Journal of RF and Microwave Computer-Aided Engineering*, vol. 30, no. 3, p. e22356, 2020.

[5] R. Rao and V. Sharma, "A comprehensive survey on machine learning techniques in antenna design and analysis," *Engineering Reports*, vol. 4, no. 3, p. e12456, 2022.

[6] S. Ali, S. I. Shah, and F. Qamar, "Machine learning in antenna design: A review," *IEEE Access*, vol. 9, pp. 10 423–10 443, 2021.

[7] N. Gupta, R. Kumar, and S. Choudhary, "Recent advances in machine learning for antenna design: State-of-the-art review," *International Journal of Electronics and Communications*, vol. 118, p. 153147, 2020.

[8] J. Xu, K. Wang, and L. Zhao, "Machine learning-assisted design and optimization of patch antennas: A comprehensive review," *IEEE Transactions on Antennas and Propagation*, vol. 71, no. 3, pp. 1845–1860, 2023.

[9] R. Garg, P. Bhartia, I. Bhal, and A. Ittipiboon, *Microstrip antenna design handbook*. Artech House, 2001.

[10] J. Mendes-Moreira, C. Soares, A. Jorge, and J. Sousa, "Ensemble approaches for regression: A survey," *ACM Computing Surveys*, vol. 45, no. 1, pp. 1–40, 2012.

[11] G. Biau and E. Scornet, "A random forest guided tour," *Test*, vol. 25, no. 2, pp. 197–227, 2016.

[12] B. M. Wilamowski, "Neural network architectures and learning algorithms," *IEEE Industrial Electronics Magazine*, vol. 3, no. 4, pp. 56–63, 2009.

[13] W. Abbas and N. Javaid, "Performance evaluation of ensemble models for wireless channel estimation," *IEEE Access*, vol. 7, pp. 67 060–67 070, 2019.

[14] M. Kaushik and S. Srivastava, "Hybrid machine learning-based models for predicting microstrip patch antenna performance," *Wireless Personal Communications*, vol. 108, pp. 187–202, 2019.

[15] A. Elmisilmani, M. Abdulrahman, A. Ibrahim, R. Hassan, and M. Alshehri, "A review of machine learning techniques for antenna design and optimization," *Journal of Microwaves, Optoelectronics and Electromagnetic Applications*, vol. 19, no. 1, pp. 18–37, 2020.

[16] S. A. Al-Gburi, S. Al-Taee, J. Qasim, M. Mhasin, T. Sahu, H. Patil, A. Soni, and V. Lazi˘c, "Machine learning techniques for antenna design and optimization: A review," *Computers, Materials & Continua*, vol. 69, no. 3, pp. 3793–3811, 2022.

[17] S. Han, Y. Tian, W. Ding, and P. Li, "Resonant frequency modeling of microstrip antenna based on deep kernel learning," *IEEE Access*, vol. 9, pp. 39 067–39 076, 2021.

[18] K. H. Ong, W. Wahab, Z. Yusof, and M. Ibrahim, "Machine learning for circular patch antenna classification based on resonance and impedance matching," *Applied Computational Electromagnetics Society Journal*, vol. 45, no. 8, pp. 1230–1240, 2025.

[19] R. Varma and J. Ghosh, "Hybrid method to predict the resonant frequencies and to characterise dual band proximity coupled microstrip antennas," in *International Journal of Electronics*, vol. 105, no. 6, 2018, pp. 910–922.

[20] F. Chen and Y.-b. Tian, "Modeling resonant frequency of rectangular microstrip antenna using cuda-based artificial neural network trained by particle swarm optimization algorithm," *Applied Computational Electromagnetics Society Journal*, vol. 36, no. 5, pp. 589–596, 2021.

[21] K. Pachori, A. Prakash, and N. Kumar, "Performance prediction of dielectric resonator based mimo antenna for sub-6.0 ghz using machine learning algorithms," *Electromagnetics*, vol. 43, no. 8, pp. 539–550, 2023.

[22] V. K. Sutrakar, A. PK, R. Bisariya, S. KK, and G. Chawan, "Design of cavity backed slotted antenna using machine learning regression model," *arXiv preprint arXiv:2502.19164*, 2025.

[23] Y. Nuray and N. Akcam, "Designing of dual band f-shaped rfid antenna using machine learning techniques," *Journal of Artificial Intelligence and Data Science*, vol. 2, no. 2, pp. 69–75, 2022.

[24] ANSYS Inc., *HFSS User's Guide*, 2020, accessed 2025.

[25] H. Li, Z. Zhang, M. Li, and Z. Zhang, "Machine learning-based design and optimization of microstrip antennas for wireless applications," *International Journal of Antennas and Propagation*, vol. 2019, pp. 1–9, 2019.

[26] X. Dong, Y.-W. Lee, and H. Yang, "Application of machine learning techniques in antenna design," *IEEE Access*, vol. 6, pp. 33 211–33 218, 2018.

[27] K. Takeda, T. Sugimoto, and H. Sato, "Performance enhancement of dual- band microstrip patch antenna using machine learning algorithms for wireless communication," *International Journal of RF and*

*Microwave Computer-Aided Engineering*, vol. 30, no. 8, p. e22394, 2020.

[28] Y. Zhang, H. Chen, S. Zhang, and T. J. Cui, “Machine learning-enabled design of high-efficiency metasurfaces,” *ACS Nano*, vol. 13, no. 8, pp. 8841–8850, 2019.

[29] T. Chen and C. Guestrin, “Xgboost: A scalable tree boosting system,” pp. 785–794, 2016.

[30] T. G. Dietterich, “Ensemble methods in machine learning,” *Multiple Classifier Systems*, pp. 1–15, 2000.

[31] F. Pedregosa, G. Varoquaux, A. Gramfort, V. Michel, B. Thirion, O. Grisel, M. Blondel, P. Prettenhofer, R. Weiss, V. Dubourg, J. Vanderplas, A. Passos, D. Cournapeau, M. Brucher, M. Perrot, and E´. Duchesnay, “Scikit-learn: Machine learning in python,” pp. 2825–2830, 2011.

[32] I. Goodfellow, Y. Bengio, and A. Courville, *Deep learning*. MIT press, 2016.

[33] Y. LeCun, Y. Bengio, and G. Hinton, “Deep learning,” *Nature*, 2015.

[34] W. Zhang, J. Wang, and H. Liu, “Deep learning-based model for predicting microstrip antenna performance,” *Electronics*, vol. 11, no. 3, p. 523, 2022.

[35] D. Opitz and R. Maclin, “Popular ensemble methods: An empirical study,” *Journal of Artificial Intelligence Research*, 1999.

[36] C. Zhang and Y. Ma, “Ensemble learning for prediction,” *Ensemble Machine Learning*, 2012.

[37] D. H. Wolpert, “Stacked generalization,” *Neural networks*, 1992.

[38] A. Shrestha and A. Mahmood, “A comprehensive review on ensemble deep learning: Opportunities and challenges,” *Information Fusion*, vol. 88, pp. 116– 138, 2022.

[39] X. Li, W. Zhang, and J. Chen, “A hybrid stacking ensemble model for predicting mechanical properties of advanced materials,” *Materials Today Communications*, vol. 36, p. 105587, 2023.

[40] L. Wang and M. Hu, “Improved ensemble learning framework for nonlinear regression problems in microwave engineering,” *IEEE Transactions on Microwave Theory and Techniques*, vol. 71, no. 2, pp. 654–667, 2023.

# 4

# Cybersecurity Threat Detection using Computational Intelligence

**Amritansh Singh[1], Monali Gulhan[2], Rashmi Sharma[3], Nitin Rakesh[4] and Gunjan Chhabra[5]**

[1,2,4]Symbiosis Institute of Technology, Nagpur Campus, Symbiosis International (Deemed University), India
[3]SVKMs NMIMS University, Mukesh Patel School of Technology Management and Engineering, Shirpur campus, Dhule, Maharashtra, India
[5]School of Science and Technology, Swami Rama Himalayan University, India

**Email:** amritanshsingh2085@gmail.com; monali.gulhane4@gmail.com; rashmi.nov30@gmail.com; nitin.rakesh@gmail.com; chhgunjan@gmail.com

## Abstract

The increasing complexity and diversity of cyber threats require the development of new detection and mitigation methodologies. This chapter discusses the integration of computational intelligence techniques, such as quantum neural networks, generative adversarial networks, artificial neural networks, and other machine learning approaches to enhance cybersecurity systems. With these emerging technologies, anomaly detection, real-time threat analysis, and risk management in IoT systems and aviation infrastructure have been significantly improved. The study points out the necessity of incorporating computational intelligence to address data scarcity, real-time processing, and adaptability to evolving threats. Case studies on quantum neural networks for virtualized environments and generative adversarial networks for SCADA system security are also highlighted. These results underscore the significance of Computational Intelligence in shaping strong, yet agile cybersecurity frameworks. Future work should therefore be focused on filling in some of the



DOI: 10.1201/9788743812739-4

existing gaps, like integration issues and ethical concerns, as well as hybrid quantum-classical models, in an attempt to further harden digital ecosystems.

## 4.1 Overview of Cybersecurity

### 4.1.1 What is cybersecurity

Cybersecurity helps you secure the systems, networks, and data from cyber-attacks. With the goal of keeping information confidential, available, and instead. Now, businesses are using digital infrastructure more, for which cybersecurity is very important. This offers several countermeasures to prevent or defend against cyberattacks, unauthorized access, and other vulnerabilities, such as ensuring the security of sensitive information.

### 4.1.2 Types of cyber threats

#### 4.1.2.1 Cyber threats

Malware (viruses, worms, and ransomware) is one of the major threats that gains access into the network to cause destruction.

- Phishing is the fraudulent practice of trying to get passwords and credit card numbers, often asked for in emails, which are usually misleading.
- Distributed denial of service (DDoS): A technique to assault some network or system in a way that causes the target to suffer from DDoS and is unable to access normally [1].

*Advanced persistent threats (APT):*

Advanced persistent threats (APT) are some of the most dangerous cyber threats, which are often characterized by stealth and long-term presence in targeted systems. Such threats involve sophisticated and targeted cyberattacks aimed at stealing data or disrupting operations over an extended period. APTs are often associated with state-sponsored groups or highly organized cybercriminals who use advanced techniques to evade detection [2].

As identified by Lee, APTs exploit vulnerabilities in systems, sustaining a persistent existence and operating unnoticed until their target is accomplished. For instance, the notorious group "APT9," believed to be of Russian origin, was reportedly responsible for several high-profile cyber-espionage campaigns. To combat APTs, organizations use intense detection mechanisms: threat intelligence, behavioral analysis, and machine learning models, according to Oluwaseun et al. [3]. Figure 4.1 shows four pillars of cyber security.

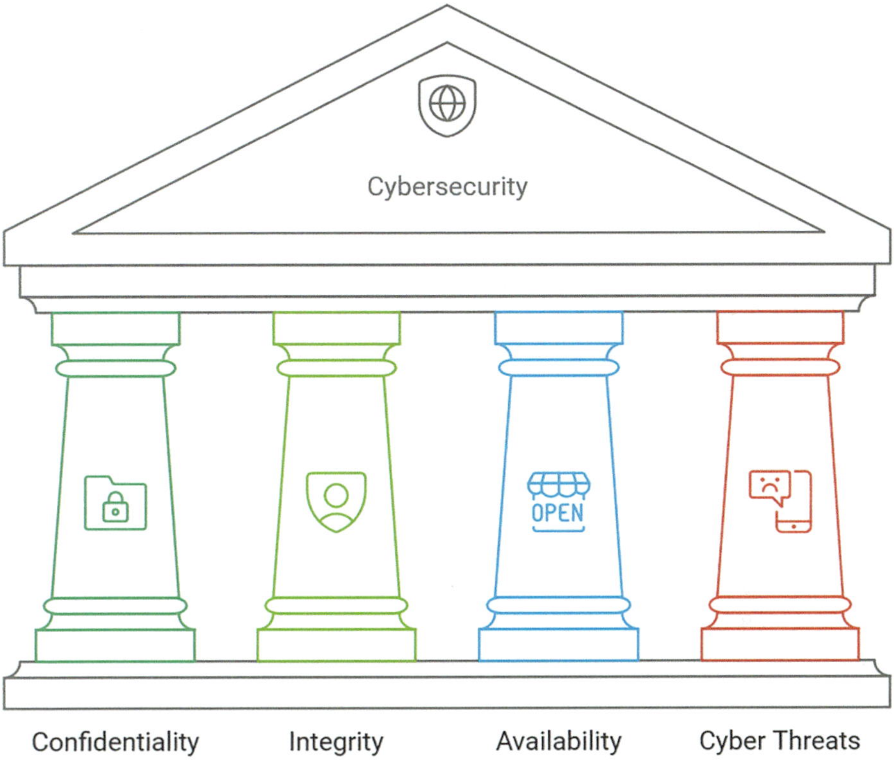


**Figure 4.1** Pillars of cybersecurity.

#### 4.1.2.2 Network security

Network security safeguards an 'organization's network infrastructure against breaches, unauthorized access, and misuse. It does it with things like firewalls, intrusion detection systems (IDS), and encryption. Network security [4] is more important than ever, with cloud computing and IoT devices increasing the amount of data that can be intercepted or exploited [11].

#### 4.1.2.3 Data security and privacy

Data security practices are guidelines and technologies we use to protect sensitive data from unauthorized access, changes, or destruction. To ensure that only the system can access the data, encryption and masking techniques are widely used for security purposes. On the other hand, Privacy makes sure that any personal data is collected and processed based on Laws like GDPR to protect users' privacy as well as comply with legal regulations [6].

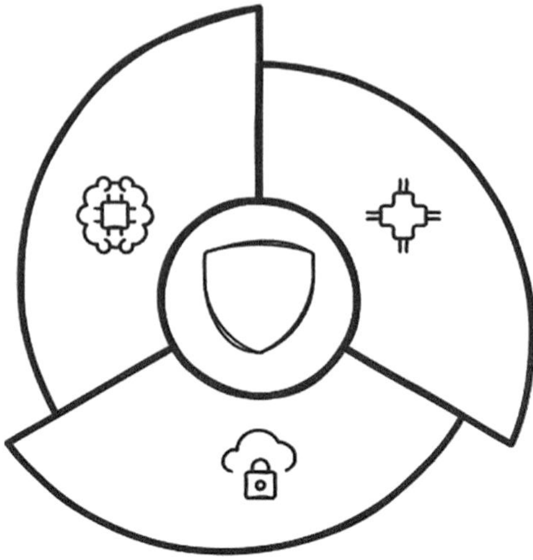


**Figure 4.2** Cybersecurity strategies.

#### 4.1.2.4 Cybersecurity and AI

The use of artificial Intelligence (AI) [7] in cybersecurity continues to grow and is being applied to deploy sophisticated methods for detecting threats, automating responses, and augmenting defenses. It looks for patterns in vast amounts of data to detect potential indicators that a cyberattack is on its way – whether the other end actually knows about it or not. These are ways you may use artificial intelligence to process data more quickly and accurately than traditional methods, and such AI is becoming a critical part of modern cybersecurity. In figure 4.2, cybersecurity strategies have been mentioned that are elaborated in next section.

## 4.2 Function of Computational Intelligence (CI) in Threat Detection

### 4.2.1 Introduction to computational intelligence (CI)

Computational intelligence (CI) is a series of technologies inspired by natural phenomena intended to solve complex issues that ordinary algorithms cannot handle. Computational intelligence (CI): Artificial neural network (ANN), fuzzy logic, genetic algorithm, swarm intelligence, etc. They are particularly valuable in cybersecurity because the ability to process enormous amounts of data and adapt to new threats is so important. CI scales defenses to mimic how the human immune system works, aiding us in intelligence-based decision-making about cyber protection.

### 4.2.2 Types of cyber threats handled by CI

CI approaches are very effective in managing various types of cyber risks, such as:

Insider threats: These occur when trusted individuals misuse their legitimate access to valuable information. A CI system can learn typical user behavior patterns and detect anomalies, which might indicate malicious activity [8].

i. Distributed Denial of Service (DDoS) Attacks: DDoS attacks flood networks with heavy traffic, which causes a disruption. Quantum neural networks (QNNs) offer faster processing speeds for real-time detection, thus allowing proactive mitigation of such attacks [2].
ii. IoT Attacks: The internet of things (IoT) devices produce varied and vast datasets. CI is critical in analyzing these datasets to identify security vulnerabilities and potential threats effectively.

The authors claim that quantum concepts and data generation can enhance anomaly detection and behavior analysis for risk management across IoT systems and aviation cyber infrastructure [9]. However, this review does not substantiate these claims with specific examples or case studies. Instead, it mainly defines QNN and other CI techniques without providing empirical evidence. Future versions might include real-world implementations, such as QNN applications in virtualized environments or GAN-based SCADA security systems, as discussed in [1] [2] and [10].

#### 4.2.2.1 CI techniques for threat detection

Several CI techniques play a vital role in enhancing cybersecurity. Figure 4.3 represents the approach of CI that can enhance the cybersecurity:

There are several CI techniques important to enhance cybersecurity that play a crucial role in the following areas [11]. Figure 4.4 mention few techniques of CI in cybersecurity:

1. **Artificial Neural Networks** can recognize patterns within data, where ANNs can indeed be quite effective in identifying hostile

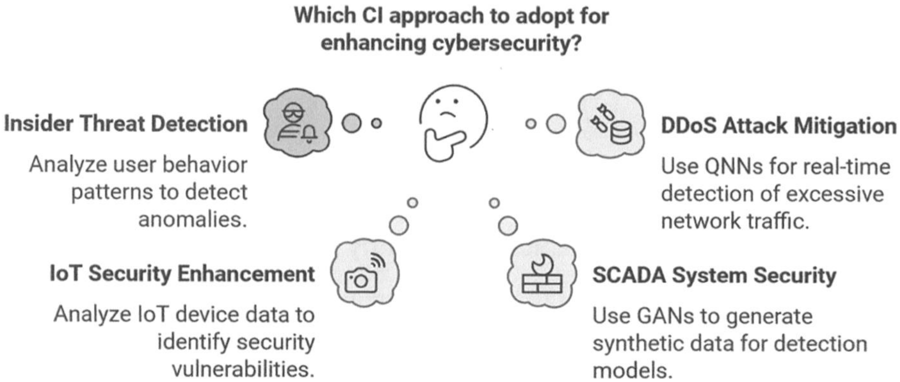


**Figure 4.3** CI approach to enhance cybersecurity.

activity or even insider threats in cybersecurity [11] [12]. They can therefore learn to discern between normal behavior and suspicious behavior if trained on past data.

2. **Fuzzy Logic:** Fuzzy logic is an approach that addresses the uncertainty and vagueness present in the data so that systems can make decisions under partial truth. It is quite apt for intrusion detection, as mostly the data is not absolutely true or false, but it is somewhere in between.

3. **Genetic Algorithms (GAs):** These optimization methods are inspired by natural selection. GAs can be used in cybersecurity for finding the best settings of intrusion detection systems or tuning model parameters, and may be used to improve the detection rate and reduce false positives.

4. **Swarm intelligence** mimics the collective behavior of social animals and could enhance danger perceived within a network, thereby enhancing response within a distributed network, such as in IoT networks [9], Devices would collaborate by sharing danger information between them to alert about anomalies.

5. **Support Vector Machine (SVM):** These are helpful for specific recognitions of cyber threats. They operate by finding the best hyperplane, distinguishing between benign and malicious data, and are invaluable in threat detection [9]. Quantum neural networks (QNNs) – a new frontier.

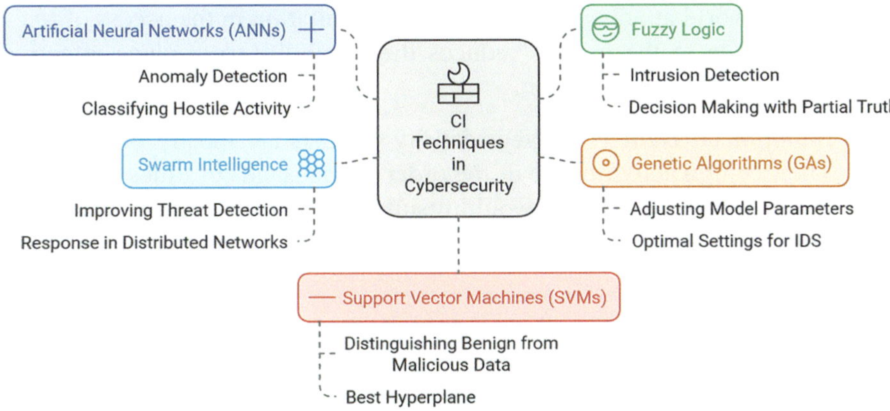


**Figure 4.4** Techniques in cybersecurity.

Quantum neural networks are one of the most radically new concepts in cyber intelligence. QNNs involve the concept of quantum computation in processing ten times faster than a standard neural network. Such a capability makes them exceptionally suitable for high-speed threat detection environments such as cloud computing and virtualized environments [13]. The authors provide the superiority of QNNs by demonstrating their capability to reduce the false positives while correctly identifying complex threats, proving the future of cybersecurity techniques to be transformed.

## 4.3 Challenges in using CI for Threat Detection

The threats in the cyber world are very dynamic, so the issues mentioned above become very fundamental in developing effective CI-based security systems. While CI may provide potential means to enhance security, there are several issues that need to be addressed to ensure effectiveness.

### 4.3.1 Data scarcity and quality

One of the biggest hurdles in deploying CI techniques is that good-quality, labelled datasets from which models are trained just are not available. Ideally, big datasets with proper labels showing whether the activity is benign or malicious are quite important for efficient machine learning [14], but difficult in cybersecurity due to many reasons:

- ***Limited Access to Real Attack Data:*** The lack of organizations that share data about cyberattacks limits the data used for training purposes in real-world attacks, and hence, such a limitation in the access of real-world data to the models reduces the capability to produce effective systems of threat detection.

- ***Imbalanced Datasets:*** Cybersecurity datasets are mostly imbalanced. Their benign activities are of the order of thousands of times more than the malicious ones. This would result in poorly performing models at detecting real threats since it becomes biased toward predicting benign activities.
- ***Synthetic Data Generation:*** GANs can be applied to fill missing data gaps; however, even when synthetic datasets are used, they may not be able to mimic all realistic threats that could occur in real environment applications, thus making the learned model less accurate.

### 4.3.2 Real-time processing and scalability

CI models must operate effectively in real-time to identify and respond to threats promptly. As organizations increasingly adopt technologies like cloud computing and the internet of things (IoT) [10], [15], [16], the volume of data generated grows exponentially, posing several challenges:

- **High Computational Load:** Many CI algorithms, such as deep neural networks, require significant computational power for both training and inference. Processing large volumes of data in real-time can strain existing resources, leading to delays in threat detection.
- **Latency Issues:** An IDS must be latency-free. Intricate algorithms may make the systems of detection latent, causing potential threats to grow before they are noticed and pursued. This is very crucial for environments that demand a prompt response.
- **Scalability in Large Networks:** With the scaling of networks in organizations, it therefore implies that scalability for CI solutions is a requirement. Traditional detection techniques may not be effective with the volumes of data associated with such large systems and thus require new techniques to ensure smooth operation [3].

### 4.3.3 Adaptability to evolving threats

Zero-day attacks are attacks on unknown vulnerabilities that make them the most difficult threat to cybersecurity. Conventional models have a hard time identifying such attacks because they have no prior knowledge of the

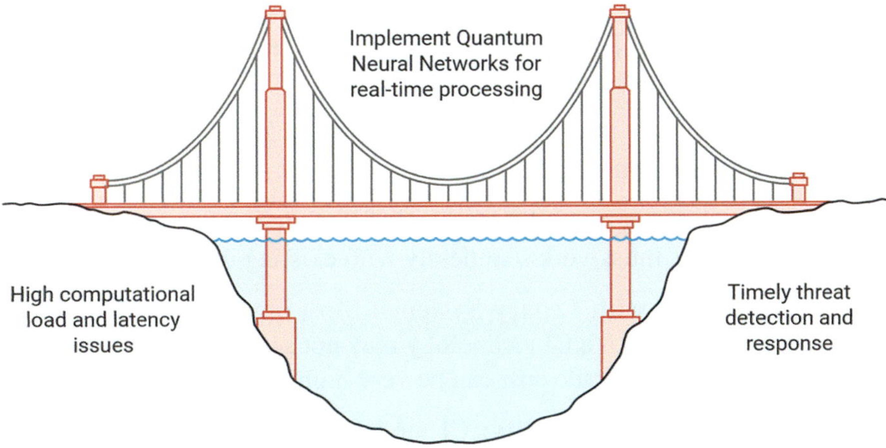


**Figure 4.5** Enhance cyber threat detection efficiency.

vulnerability. With the reference of figure 4.5, few cyber threat detection techniques are discussed here. Few examples of zero-day attack followed by techniques to handle it are discussed as follows:

1. Stuxnet (2010): A complex worm that attacked industrial control systems, specifically Siemens PLCs, used in Iran's nuclear facilities. It exploited multiple zero-day vulnerabilities and was believed to be state-sponsored.
2. Google Chrome Zero-Day (2022): A vulnerability in the Chrome browser (CVE-2022-1096) allowed attackers to execute arbitrary code, which was quickly patched by Google after its discovery.

Recent approaches to tackling zero-day attacks:

1. AI-Powered Threat Detection: Machine learning models analyze network behavior and identify anomalies that may point to zero-day exploits. Techniques such as random forest-SMOTE help in rare pattern detection and unseen patterns.
2. Threat Intelligence Sharing: ISACs are platforms for organizations to share data about zero-day vulnerabilities, thus collectively enhancing defenses.
3. Quantum Neural Networks (QNNs): Building upon quantum computing, QNNs enhance pattern identification concerning unusual, zero-day-type patterns that point toward attacks.

4. Behavioral Analysis: This system would look at what is happening from application behavior in case of irregular activities, and hence a deviation that represents exploitation by definition.

### 4.3.4 Integration and interoperability

As organizations deploy multiple cybersecurity solutions, integration becomes critical. CI systems must work seamlessly with existing infrastructure:

- **Compatibility with Legacy Systems:** Most organizations have older systems with which a CI technology may not seamlessly integrate. The replacement or upgrade cost can be very high and take a long time.
- **Data Privacy and Security:** CI models require massive amounts of data, which subjects them to criticism about privacy. Organizations have to ensure that the private information is protected while being collected and analyzed.
- **Human-Computer Interaction:** Security professionals need to interpret the results produced by CI models. The complex algorithms can work as "black boxes," which means that often, the processes that lead to certain decisions cannot be understood by analysts.

## 4.4 Applications of CI in Cybersecurity

CI has gained considerable advancements in many areas of cybersecurity and now brings novel solutions for detection, analytics, and mitigation of cyber threats. Its ability to handle large data flows and evolve with shifting attack patterns makes CI an integral component for digital space protection. In the following section, several prominent applications of CI in cybersecurity are discussed [17].

### 4.4.1 Quantum neural networks (QNN) applied to virtual cyber threat detection

Quantum neural networks (QNNs) are an innovative application of CI to cybersecurity. Because organizations are increasingly moving to cloud infrastructures and virtual environments, it has become increasingly difficult to monitor these systems and protect them from malicious attacks [18]. Traditional forms of security do not work as well with the volume and complexity of data occurring within these environments (Figure 4.6).

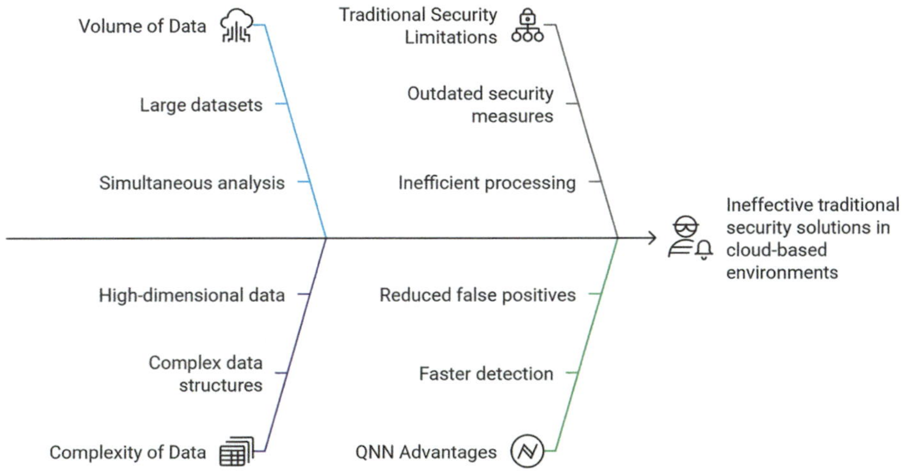


**Figure 4.6** Enhancing cybersecurity and QNN.

QNNs apply quantum principles in computing to enable faster information processing compared to classical neural nets. QNNs will process big data in parallel through the use of qubits, which facilitates the network's processing huge amounts of data within a much shorter period and highly increased accuracy rates concerning anomaly detection. QNNs minimize false positives in the identification of threats like distributed denial-of-service (DDoS) or advanced persistent threat (APTs), as seen in the research paper.

**Real-world implementation:** The quantum computing techniques are penetrating into the security frameworks of cloud service providers like Amazon Web Services and Google Cloud, as their threat detection capabilities are being heightened.

### 4.4.2 SCADA system security using generative adversarial networks (GANs) [19]

**Supervisory control and data acquisition systems** play a key role in managing critical infrastructures, including energy, water, and manufacturing. However, these systems remain highly vulnerable to cyber-attacks because their traditional security models lack adequate training data.

GANs represent a new approach with respect to the generation of artificial data that mimics real network traffic, making it easier to train security models. The paper explains how GANs can be used to improve threat detection in SCADA environments through the creation of realistic training scenarios for the model.

**Real-world implementation:** After the high-profile cyberattacks on infrastructure, organizations began using GANs to enhance their capability to detect attacks effectively, as they made sure they could better identify anomalies in system behavior.

### 4.4.3 Efficiency of advanced detection approaches

The use of ANN autoencoders has been highly effective for insider threat detection. The systems learn normal patterns of user behavior and flag anomalies, achieving a detection accuracy of 94.3%, outperforming significantly. Traditional rule-based systems. In fact, as Abrahams et al. in [2] have proven, ANN-based systems reduce false positives while increasing the detection rate for malicious activities.

Another comparable development is Quantum Neural Networks, which rely on quantum principles for fast data processing and anomaly detection. QNNs have exhibited higher accuracy for Distributed Denial-of-Service (DDoS) and Advanced Persistent Threats (APTs) detection with fewer false positives compared to other networks.

Generative adversarial networks (GANs) also contribute significantly by generating synthetic datasets for training, which enhance the robustness of threat detection systems, especially in critical infrastructure security like SCADA systems.

### 4.4.4 Machine learning and network analysis for IoT and CPS security

As internet of things and cyber-physical systems become increasingly ubiquitous in daily life, solutions to secure these technologies become only more urgent. Such systems produce massive volumes of data, providing an incredibly rich environment for cyber threats.

Techniques like Random Forest-SMOTE, for example, are also used for anomaly pattern detection over IoT networks. Because they handle class imbalances, they increase the chances of the model reporting malicious activity.

**Real-world implementation:** Smart city initiatives are using CI-based solutions to protect interconnected devices and systems. The machine learning models do the real-time analysis of network traffic, so any suspicious behavior is detected. This further improves the security posture of the urban infrastructure.

### 4.4.5 AI and cyber threat intelligence (CTI) for cyber supply chain security

The more connected the world gets to organizations, the higher the priority security has become in cyber supply chains [11]. CI is now merged with Cyber Threat Intelligence (CTI) for predicting and mitigating cyber threats along the supply chain.

By applying machine learning algorithms [20] with the information available through CTI, organizations are able to discover and start addressing their supply chains for vulnerabilities and potential threats. The research study here emphasizes how this method leads to boosting the detection rates as well as responses to threats.

**Real-world implementation** after incidents such as SolarWinds breach: Companies have begun to implement AI-based security in the supply chain to predict and react to threats before they impact operations.

### 4.4.6 Artificial immune systems for aviation cybersecurity

The safety and efficiency measures for aviation systems require protection against cyber-physical systems (CPS). Artificial immune systems (AIS) have been applied and hence been implemented to detect false data injection (FDI) attacks, as some attacks that manipulate important flight data.

In AIS, the negative selection algorithm identifies normal behavior and flags deviations as potential threats. Using such a methodology coupled with predictive models will help preserve integrity in the aviation systems.

**Real-world implementation:** Aviation organizations are increasingly adopting AIS to monitor flight systems, such that any unauthorized data manipulation is detected and mitigated swiftly, thus maintaining operational safety.

## 4.5 Case Studies and Real-World Applications of CI in Cybersecurity

The practical applications of CI in cybersecurity are diverse and numerous, as there is a need for new approaches to overcoming increasingly sophisticated cyber threats. Here are a few cases of key studies demonstrating the efficiency of CI techniques in real-life conditions.

### 4.5.1 Quantum neural networks for cyber threat detection in virtualized environments

#### Case Study: Cloud Service Providers

Thus, securing virtual environments against distributed denial of service and advanced persistent threats while associated businesses are being transferred to cloud infrastructures is a great challenge. Traditional security solutions are not easily capable of dealing with the scale and complexity of data generated within these environments.

Thus, QNNs, in turn, provide an innovative approach to the scenario. Quantum computational principles enable QNNs to handle sets of large-scale data in parallel, which would quickly flag anomalies and security threats. This paper emphasizes their capabilities of drastically minimizing false positives and improving the detection accuracy with respect to classical methods.

**Real-world application:** Major cloud service providers, such as Amazon Web Services and Microsoft Azure, are researching QNNs to enhance their security frameworks, making it feasible for them to keep watch over and prevent more complex threats.

### 4.5.2 Generative adversarial networks (GANs) for SCADA system security

#### Case Study: Critical Infrastructure Protection

Supervisory control and data acquisition (SCADA) systems are crucial because they have been essential in the operation of critical infrastructures such as power grids and water treatment facilities. Because of these systems' vulnerabilities to cyberattacks, security should thus be tightened.

GANs have come as a solution that generates synthetic data to supplement limited datasets used in the training of threat-detection models. GANs can simulate realistic network traffic scenarios that improve the training of algorithms in machine learning, hence leading to better detection of anomalies [21].

**Real-world implementation:** As recent as the 2015 Ukraine power grid attack has inspired most of the utility companies to go for GAN-based approaches that can sharpen the SCADA system security, so they can easily detect and respond to cyber threats.

### 4.5.3 Insider threat detection using ANN autoencoders

#### Case Study: Banks

Insider threats are probably the most damaging of them all, especially where sensitive data is concerned, in finance, as well as healthcare. All these require advanced analytical capability.

It can learn normal patterns using artificial neural network (ANN) autoencoders to identify anomalies in user behavior patterns. Autoencoders could reach an amazing detection accuracy of 94.3% with regard to latent insider threats, far outperforming traditional approaches.

**Practical implementation:** Large banks and healthcare organizations are using ANN autoencoders to learn the patterns that characterize employee activity. This helps identify any unauthorized access, where access to sensitive information occurs quickly.

### 4.5.4 Machine learning and network analysis for IoT and CPS security

#### Case Study: Smart Cities

Thereby, with the proliferation of IoT devices in smart cities, the security of those interconnected systems is vital. The voluminous data being generated by the devices could be too daunting to handle for security measures based on traditional back-end architectures.

Hence, various techniques are applied to IoT anomaly detection, such as the random forest-synthetic minority over-sampling technique (SMOTE). These techniques fight class imbalances in the dataset, thus boosting the chance of malign activity identification.

**Real-world implementation:** Cities embracing smart city technologies are using the power of machine learning models to monitor data from traffic and utilities in real time. Such abilities enable cities to move ahead of cyber threats in real time [22] so that municipal systems can be proactive about protecting their critical infrastructures.

### 4.5.5 Cyber Supply Chain Security with AI and CTI

#### Case Study: Global Manufacturing Companies

Within an organization that is increasingly connected, CSCs are prime targets for cybercrime. The ravages can be severe, touching many stakeholders in a supply chain.

This integrated AI system with Cyber Threat Intelligence (CTI) will allow the proactive identification of vulnerabilities and potential threats within the organization's supply chain. It will increase the detection rate, and it will also help the organizations to effectively respond to new emerging threats.

**Real-world implementation:** With high-profile supply chain attacks, companies in manufacturing, technology, etc., have initiated AI-driven security measures that analyze CTI data as the basis to anticipate and mitigate risks spread across their supply chains, leading to greater resilience against cyber threats.

### 4.5.6 Artificial immune systems for aviation cybersecurity

#### Case Study: Aviation Safety

Critical to the success of CPS is ensuring that in the aviation industry, they do not malfunction, as they do in other sectors, and somehow continue to operate safely and efficiently. FDI attacks by cyber threats could easily manipulate critical flight data, resulting in catastrophic failures.

Artificial immune systems AIS apply concepts from the biological immune system, which is capable of detecting anomalies within network behaviors. Applying a negative selection algorithm can detect anomalies by patterns, and this way, it can increase security for aviation systems.

**Real-world implementation:** More and more aviation organizations are seeking the implementation of AIS so their networks can be monitored for the integrity of flight data and such manipulations, to ensure that they are detected before they may impact operations.

## 4.6 Future Directions for CI in Cyber Threat Detection

Future advances in computational intelligence (CI) in cybersecurity: As the whole world is facing problems with cyber threats, several organizations are competing for an innovative solution to challenge the ever-evolving landscape of cyber threats. The attackers are becoming increasingly sophisticated, and data generation is rising sharply; hence, CI should evolve to enhance its capabilities for threat detection and mitigation [23] [24]. Some of the crucial future directions for CI in cybersecurity are mentioned below:

### 4.6.1 Real-time threat detection and decision-making

Real-time detection of threats is one of the most crucial challenges to CI [4] [25-27]. The amount of data that is being generated with the proliferation of cloud computing, IoT, and smart technologies calls for the need for immediate detection and neutralization of threats.

- **Edge Computing:** Edge computing would cut down the latency since it processes data nearer the source; threat response times for it would also improve. Real-time CI model analysis of the data will enable a more rapid response to possible threats.
- **Federated Learning:** This is the strategy that lets the aggregation of learning from multiple devices toward a common model while keeping data localized. Federated learning [28] enhances real-time detection capability without the compromise of data privacy, hence allowing CI systems to update their threat detection models continually.

### 4.6.2 Enhancing scalability in large, distributed networks

The other important issue for the CI model when organizations extend their digital infrastructures is scalability. The biggest challenge here would be to ensure that the cybersecurity solutions are able to handle large amounts of data in a distributed setup.

- **Cloud-Native CI Architectures:** Inherently cloud-native CI models will make them scale efficiently with the growth of data. The architectures are going to be dynamic by nature and can allocate resources based on demand, so they are fair and consistent across very wide networks.
- **AI-Based Network Optimization:** It will be possible to locate the most important nodes and connections in the networks through the use of AI in future CI systems, thus giving a basis for targeted security measures. Resource optimization will facilitate proper allocation, enhancing the scalability of the cybersecurity solutions as a whole [29].

### 4.6.3 Adaptive learning for emerging threats

With the speed at which cyber threats are changing, CI systems must be constantly evolving and learning adaptive capabilities from new data streams being introduced all the time. Adaptability in an approach like this is crucial for zero-day attacks and other emerging threats.

- **Continuous Learning Models:** Future CI systems would embed models that leverage it to learn in real time. That means such models may be updated in real time to fit new attack patterns, as threats evolve, so detection must be effective.

- **Transfer Learning:** Using knowledge from one domain to improve performance in another, transfer learning will allow CI models to generalize over several kinds of threats, reducing the magnitude of large datasets that would otherwise be needed to build the model for every kind of new attack.

### 4.6.4 Integration with quantum computing toward enhanced cybersecurity

Quantum computing holds the potential to transform CI in cybersecurity due to its capability to providing unprecedented computational powers. With this quantum technology maturing, it can be integrated into CI frameworks and enhance threat detection.

- **Hybrid Quantum-Classical Models:** It would enable organizations to leverage the strengths of both approaches because systems that combine the classical CI methods with quantum algorithms are being developed. The hybrid model would also improve capabilities for tasks like anomaly detection and cryptographic security.
- **Quantum-Resistant Algorithms:** With the widespread use of quantum computing across the globe, it will catalyze a need for CI systems that offer integrity against quantum attacks. Algorithms that can resist the existing threats from the adversaries enabled by quantum computing would be highly important for future deployments of any cybersecurity measure.

### 4.6.5 Ethical AI and Explainability in Cybersecurity

The integration of CI systems in cybersecurity and their enhanced integration make ethical AI and transparency in the decision-making process a focus issue increasingly.

- **Explainable AI (XAI):** Support for explainability will be a minimum requirement in future CI models. Techniques for XAI will enable cybersecurity experts to understand the reasoning behind a decision made by the model, which will then inform decisions and allow appropriate action.
- **CI Ethical Standards:** Organizations need to establish clear ethical standards for the application of CI in cybersecurity, with AI-driven decisions unwittingly causing harm or discriminating. Included issues

are data privacy, bias in algorithm development, and even misuse of AI technologies.

### 4.6.6 Cyber threat intelligence integration for proactive security

Integrating CI with Cyber Threat Intelligence (CTI) would be a foundational step in the development of proactive cybersecurity measures. Combining real-time threat intelligence with CI capabilities can strengthen an organization's defenses dramatically. Figure 4.7 is the pictorial representation of enhancing cybersecurity with advanced technology.

- **Proactive CI Systems:** The above information regarding CTI from the data it collects will allow the CI models to predict emerging threats and set the parameters for the detection of such emergent threats. Hence, it would make a proactive system rather than a reactive one, keeping the organizations ahead of the emergent threats rather than waiting for them to emerge.
- **CTI-Informed Automated Responses:** The CI systems of the future are most likely to embrace and even automate their responses based on the analysis conducted by CTI. The system would then initiate predefined steps like isolating infected devices or applying patches on mere detection, thus cutting down the time between detection and mitigation.

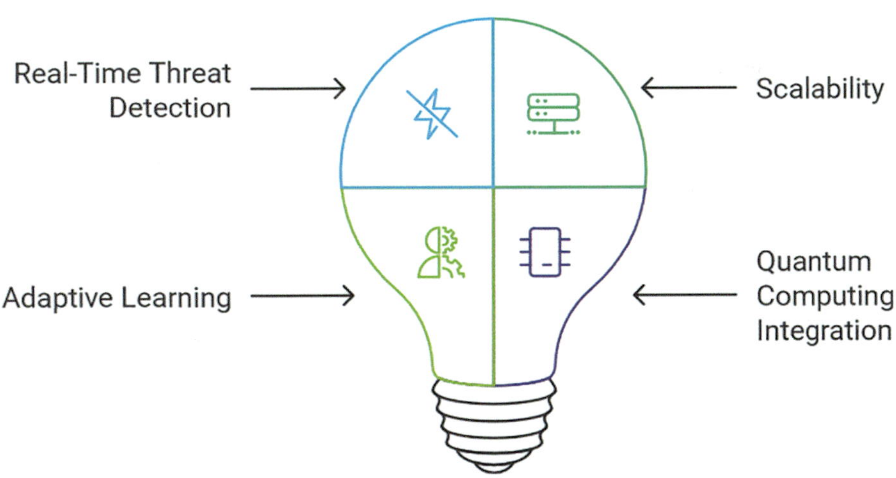


**Figure 4.7** Enhancing cybersecurity with advanced technologies.

### 4.6.7 Novel frontiers in cybersecurity threat detection

In addressing the redundancy criticism and the perceived lack of innovation, the following avenues open up novel opportunities to advance the area of cybersecurity threat detection with computational intelligence (CI):

1. **AI-driven deception systems**

   Cybersecurity systems can evolve to integrate AI-powered deception technologies, such as honeypots enhanced with generative AI, capable of creating adaptive, dynamic bait environments that confuse and mislead attackers. Unlike static honeypots, these systems would evolve in real-time, imitating high-value assets and recording malicious tactics for threat intelligence.

2. **Neuro-symbolic AI for cybersecurity**

   A promising direction is the integration of symbolic reasoning with neural networks, combining logical inference with learning-based techniques. Neuro-symbolic AI can improve interpretability in cybersecurity systems, allowing for real-time decision-making that is both explainable and effective against zero-day threats [30].

3. **Adaptive quantum-cyber threat models**

   While quantum computing exhibits promise, this integration with CI can go beyond that of the application currently. Therefore, the simulation and countering of quantum-based cyber-attacks will become significant as quantum computing becomes mainstream when developing quantum threat models. Hybrids can potentially optimize both the classical and the quantum computations into superior anomaly detection.

4. **Emotion-aware threat detection for social engineering attacks**

   Explore the application of NLP and sentiment analysis to detect emotion-based manipulations in phishing and other social engineering tactics. This layer of defense would analyze communication tone, emotional triggers, and behavioral patterns to identify and mitigate threats before exploitation [6].

5. **Blockchain-integrated threat analysis systems**

   Blockchain technology can be used to create immutable, decentralized threat intelligence-sharing platforms across organizations. Smart contracts can be used to automate response protocols across interconnected systems, thereby ensuring faster and more coordinated actions against large-scale attacks [31] [32].

6. **Sustainability in CI models for cybersecurity**

   With the increasing energy requirements of CI models, energy-efficient algorithms and hardware optimized for cybersecurity will be a subject for future research. Green cybersecurity solutions would help cater to the increased demand for sustainable digital infrastructure protection [33-35].

7. **Personalized CI models for IoT devices**

   To enhance the security of IoT, future CI systems should personalize threat detection mechanisms based on the behavior of the device and the user profiles. Lightweight, device-specific models will provide efficient, localized security without overburdening the central systems [36-38].

Focusing on these uncharted areas, this paper sets the ground for researchers and practitioners to innovate beyond current limitations and address emerging challenges in cybersecurity [39][40].

## 4.7 Conclusion

This chapter pushes forward the transformative potential of computational intelligence in addressing the growing complexities of cyber threats. Through concepts such as quantum neural networks (QNNs), generative adversarial networks (GANs), and artificial neural networks (ANNs), computational intelligence has been shown to detect anomalies with the aim of mitigating risks, thus automating responses in critical domains such as IoT systems, SCADA infrastructure, and aviation security. These include large-scale networks with proactive anomaly management, enhanced insider threat detection, and case studies of practical applications.

However, there are still significant challenges. The dependency on high-quality labelled data is a bottleneck, especially for imbalanced datasets that are common in cybersecurity. The computational demands of CI algorithms lead to scalability and latency issues in real-time threat detection. Ethical concerns such as data privacy and the lack of explainability in CI models further hinder trust and adoption. Integration with legacy systems also creates difficulties in the seamless deployment of modern cybersecurity solutions.

To overcome the mentioned limitations, the chapter opens new avenues of research. They are AI-based deception systems, neurosymbolic AI for increased explainability, and emotion-aware detection mechanisms for countering social engineering attacks. New ideas about blockchain-based threat intelligence sharing and adaptive quantum-cyber threat models pave the way toward more secure and efficient systems. Sustainable CI models and personalized IoT security solutions will ensure the scalability and relevance in different environments.

In conclusion, while computational intelligence has shown great promise in strengthening cybersecurity, addressing the current limitations and exploring these novel frontiers will be critical for advancing the field. This work aims to inspire researchers and practitioners to innovate beyond existing paradigms, ensuring the resilience of digital ecosystems against evolving cyber threats.

## References

[1] Liam O'Connor, Emma Burke. "From Heuristics to AI: The Evolution of Malware Detection Techniques." Journal of Computer Threat Studies, 2023.

[2] Mauro José Pappaterra, Francesco Flammini. "Bayesian Networks for Online Cybersecurity Threat Detection." Studies in Computational Intelligence, 15 December 2020.

[3] Jonas Smith, Ines Rodrigues. "Leveraging AI for Advanced Persistent Threat Detection in Government Networks." Public Sector Cybersecurity Review, 2023.

[4] Ahmed Farooq, F. Rizwan. "Cybersecurity Automation: The Role of Artificial Intelligence in Real-Time Threat Prevention." Cyber Risk and Resilience Journal, 2023.

[5] Ijiga, O. M., Idoko, I. P., Ebiega, G. I., Olajide, F. I., Olatunde, T. I., & Ukaegbu, C. (2024). Harnessing adversarial machine learning for advanced threat detection: AI-driven strategies in cybersecurity risk assessment and fraud prevention.

[6] Oliver Roberts. "The Use of Natural Language Processing in Threat Detection Systems." AI and Linguistic Security Journal, 2023.

[7] W. S. Ismail (2024). "Threat Detection and Response Using AI and NLP in Cybersecurity." Journal of Internet Services and Information Security (JISIS), 14(1), 195-205. DOI: 10.58346/JISIS.2024.I1.013

[8] Manoharan, A., & Sarker, M. (2023). Revolutionizing Cybersecurity: Unleashing the Power of Artificial Intelligence and Machine Learning

for Next-Generation Threat Detection. DOI: https://www. doi. org/ 10.56726/IRJMETS32644, 1.

[9] Kai Lin, Wei Zhang. "Real-Time AI for IoT Cybersecurity Threat Detection: A Case Study." Journal of Internet of Things Engineering, 2024.

[10] Jaspreet Singh, Nidhi Khurana. "Deep Learning Techniques for Anomaly-Based Threat Detection in Cyber Systems." Cybernetics and Systems Engineering Journal, 2023.

[11] Ojo, B., & Aghaunor, C. T. (2024). AI-driven cybersecurity solutions for real-time threat detection in critical infrastructure. International Journal of Science and Research Archive, 12(02), 1716-1726.

[12] Basiru A. Olafuyi (2023). "Artificial Intelligence in Cybersecurity: Enhancing Threat Detection and Mitigation." International Journal of Scientific and Research Publications, 13(12), 194-200.

[13] M. Kowalski, N. Ahmed. "Threat Detection in Cloud Environments Using AI: A Review of Emerging Techniques and Tools." Journal of Cloud Computing Advances, 2024.

[14] Santosh Gore, Sajeesh Hamsa, Sujata Roychowdhury, Gayatri Patil, Sujata Gore, S. Karmode. "Augmented Intelligence in Machine Learning for Cybersecurity: Enhancing Threat Detection and Human-Machine Collaboration." 2023 Second International Conference on Augmented Intelligence and Sustainable Systems (ICAISS), 638-644.

[15] Fadwa M. Alrowais, Sami Althahabi, Saud S. Alotaibi, Abdullah Mohamed, M. A. Hamza, Radwa Marzouk (2024), "Automated Machine Learning Enabled Cybersecurity Threat Detection in Internet of Things Environment." Computer Systems Science and Engineering, 45(1).

[16] Ullah, F., Naeem, H., Jabbar, S., Khalid, S., Latif, M. A., Al-Turjman, F., & Mostarda, L. (2019). Cyber security threats detection in internet of things using deep learning approach. IEEE access, 7, 124379-124389.

[17] William Zhang, Nathaniel Bryce, Naomi Wingate. "A Survey on the Application of Reinforcement Learning in Threat Detection for Cybersecurity." Journal of Artificial Intelligence Research, 2024.

[18] Fatima Rashed Alzaabi, Abid Mehmood. "A Review of Recent Advances, Challenges, and Opportunities in Malicious Insider Threat Detection Using Machine Learning Methods." IEEE Access, 2024.

[19] Temitayo Oluwaseun Abrahams, U. Okoli, Ogugua Chimezie Obi, Adebunmi Okechukwu Adewusi. "Machine Learning in Cybersecurity: A Review of Threat Detection and Defense Mechanisms." World Journal of Advanced Research and Reviews, 30 January 2024.

[20] Taher Al-Shehari, D. Rosaci, R. Nawaz. “Enhancing Insider Threat Detection in Imbalanced Cybersecurity Settings Using the Density-Based Local Outlier Factor Algorithm.” IEEE Access, 2024.

[21] Anita Deshmukh. “AI-Augmented Cybersecurity Training: Bridging the Gap in Threat Detection Skills.” Journal of Education in Cybersecurity, 2023.

[22] Usa Raja Kumar Kohli, Punit Goel. “Cybersecurity Threat Detection Using Machine Learning: Developing Models for Detection of Cybersecurity Threats in Real Time.” Independent Researcher, 2024.

[23] Reyadh Alluhaibi (2024) “Quantum Machine Learning for Advanced Threat Detection in Cybersecurity.” International Journal of Safety and Security, International Journal of Safety & Security Engineering, 14(3).

[24] Labu, M. R., & Ahammed, M. F. (2024). Next-Generation cyber threat detection and mitigation strategies: a focus on artificial intelligence and machine learning. Journal of Computer Science and Technology Studies, 6(1), 179-188.

[25] Nina Patel. “AI Ethics in Cybersecurity: Challenges in Automated Threat Detection and Decision Making.” Ethics in Artificial Intelligence Journal, 2024.

[26] George Matthews. “Zero-Day Threat Detection Using AI: Overcoming Challenges in Predictive Cybersecurity.” Journal of Cyber Threat Intelligence, 2023.

[27] F. H. Ali, K. Parker. “The Impact of AI on Cybersecurity Workforce Dynamics: Enhancing Threat Detection Capabilities.” Journal of Cyber Workforce Studies, 2024.

[28] Sara Salim, Nour Moustafa, Mohamed Hassanian, David Ormod, J. Slay. “Deep-Federated-Learning-Based Threat Detection Model for Extreme Satellite Communications.” IEEE Internet of Things Journal, 1 February 2024.

[29] Hiroshi Tanaka, Yuki Kato. “AI-Powered Threat Detection: Optimizing Cybersecurity Systems Using Hybrid Models.” International Journal of Emerging Technologies, 2023.

[30] Francesco Bergadano, G. Giacinto. “AI for Cybersecurity: Robust Models for Authentication, Threat and Anomaly Detection.” Algorithms, 7 July 2023.

[31] Faisal A. Mahmood, Kareem A. Hamid. “Federated Learning in Cybersecurity: A Distributed Approach for Threat Analysis and Detection.” Journal of Distributed Computing, 2024.

[32] Raman Mehta. “Blockchain-Integrated AI Systems for Enhanced Cybersecurity Threat Detection.” Journal of Secure Systems, 2024.

[33] Lisa Morgan. "AI-Powered Security Information and Event Management Systems for Threat Detection." Journal of Cyber Operations Research, 2024.

[34] Benjamin Carter. "Advances in Intrusion Detection Systems Through Machine Learning Algorithms." Network Security Innovations Journal, 2023.

[35] Rebecca Taylor. "The Future of AI in Cybersecurity: A Focus on Proactive Threat Detection." Journal of Emerging Security Technologies, 2023.

[36] Heyao Chen, Zepeng Shen, Yong Wang, Hu Ke, Jian Xu. "Threat Detection Driven by Artificial Intelligence: Enhancing Cybersecurity with Machine Learning Algorithms." World Journal of Innovation and Modern Technology, 18 November 2024.

[37] Mohammed Rizvi. "Enhancing Cybersecurity: The Power of Artificial Intelligence in Threat Detection and Prevention International Journal of Advanced Engineering Research and Science, 10(5), 055-060.

[38] Sakthiswaran Rangaraju (2023), "AI Sentry: Reinventing Cybersecurity Through Intelligent Threat Detection." EPH - International Journal of Science and Engineering, 9(3), 30-35.

[39] Basiru A. Olafuyi (2023). "Artificial Intelligence in Cybersecurity: Enhancing Threat Detection and Mitigation." International Journal of Scientific and Research Publications, 13(12), 194-200.

[40] Sarah J. Thomsen, Roger P. Davies. "Machine Learning for Network Security: Evolving Threat Detection Capabilities in Large-Scale Infrastructures." Journal of Network Security Studies, 2023.

# 5

# Optimizing Agile Software Development Process: A Focus on User Story Splitting Techniques

**Bhawnesh Kumar[1], Dinesh Chandra Dobhal[2], Tinku Singh[3], Ishwari Singh Rajput[4], Ashwani Kumar[5] and Harendra Singh Negi[6]**

[1,2,6]Department of Computer Science and Engineering,
Graphic Era Deemed to be University, India
[3]School of Information and Communication Engineering,
Chungbuk National University Cheongju-si, Republic of Korea
[4]School of Computing, Graphic Era Hill University, India
[5]Department of Computer Science and Engineering,
Shri Ram Group of Colleges, India

**Email:** bhawneshmca@gmail.com; dineshdobhal@gmail.com;
tinku.singh@cbnu.ac.kr; rajput.ishwari@gmail.com;
getashwanikumar@gmail.com; mail.harendrasinghnegi@gmail.com

## Abstract

User story is the textual format of requirements in Agile Software Development. There are three sections in a user story which represent who, what, and why, meaning who is going to perform, what is happening, why need to do. This is the way of the requirement presented in that form. Many works have been done for large user stories where the requirement can have a high length to implement, but that shows the complexity. To break down the larger user story into smaller ones so that implementation can be done easily. The development team of agile software improved their performance if the user story is clear to everyone. To split the single user story into many with the aid of splitting techniques are presented. But user story splitting is a very challenging task. A large user story will possibly impact planning, sizing, review, and velocity of the in-progress work. Smaller user stories enable faster, more frequent, and



DOI: 10.1201/9788743812739-5

consistent iterative development. It also reduces the risk of failure in achieving the sprint goals. In this work, the focus is on user story splitting techniques that optimize the Agile development process. Learning to split large user stories into smaller user stories is one of the best things a team can do to improve their scrum workflow. Therefore, splitting improves the team's performance in terms of completion rate, sprint velocity, defects, and cycle time. The contribution of our work is to break the complex user story into small user stories so that implementation can be done easily, identification and understanding requirements, and reduce the development time of Agile software, and also give the research directions of user story splitting.

## 5.1 Introduction

Complex or large user stories mislead the developer who takes the time to implement, and also affects the effort, that is, the reason software does not complete on time [1]. There is a need to split the large or complex user story into smaller user stories so that they are easily implemented. It is typical practice when dealing with user stories to split them into tasks [2]. However, splitting can be quite difficult to determine if and how a tale should be divided. A poorly crafted user story has the potential to mislead the developer [3], or one that depends too much on others may take too long and cause delays, missing deadlines. which will lead to an unhappy client and extra development costs due to the longer development process. Splitting can be done by various methods to improve the Agile software performance. It has its positive and negative impacts on the development of Agile software. The slicing of the user story has two types: horizontal and vertical [4]; it will be used platform-wise. Splitting can be performed on behalf of roles, types of users, functional and non-functional requirements, and may be many more. But the type of splitting method used depends on the requirements of Agile software. Splitting allows the team to improve their performance in completing their task on time. The development team can give good feedback [5], risks are also reduced, and overall, there is an efficiency improvement. The splitting is required to manage the user stories, deliver faster, estimation better, reduce the risks, and increase flexibility [6]. Story splitting user stories avoids overwhelm by giving the team smaller, more manageable pieces of work. Story splitting can be a powerful tool for Agile teams to increase productivity, focus on high-value assets, and minimize the risk of bugs [7]. Natural language processing (NLP) can also help to split the user stories, which is an automatic data-driven approach, highlighting the integration of artificial intelligence (AI) [8]. The contribution of our work is to break the complex user story into small user stories so that implementation can be done easily, identification and understanding requirements, and reduce the development time of Agile software.

This chapter is focused on the splitting techniques for the user story to enhance the performance of the Agile development team. Possible research gaps are identified by the systematic literature review, discussion of vertical and horizontal slicing, pros and cons of different types of splitting techniques are also examined with given suitable examples, along with the impact of user story splitting, and also provides research directions.

## 5.2 The Significance and Challenges of User Story Splitting

Agile team of Scrum methodology focused to implement the user stories [9]. The user story should be created fine, planned, prioritized, and estimated. The backlog has the all-user stories which are going to be assigned to the development team for every iteration [10]. Each iteration compiled the assigned user stories and left non-completed user stories, which are again assigned in the next iteration. Maybe the large user story should not be completed within iteration [11], which is why large user stories should be split into smaller ones to complete within an iteration. This is the significance of user story splitting [12]. The splitting of user stories helps to maintain a consistent flow of value delivery, make it easier for teams to estimate work, and facilitate quicker feedback loops [13].

Which user story needs splitting is also a challenge in ASD, and under which method of splitting. Appropriate criteria for splitting the user stories are also the challenge, which has pitfalls of ASD [14]. To overcome this pitfall, Agile teams must understand that user stories are requirements that evolve with the Agile project [15]. Each user story requires its unique technique to split, depending on its complexity, the knowledge of the team, and the project context. The various splitting techniques described in this book chapter provide a robust starting point. This chapter also discusses the possible research directions toward user story splitting.

## 5.3 Literature Review

The following related work is focused on splitting user stories in Agile software.

Liskin et al. [16] explained the granularity concept for user stories to reduce the scope so that quality and clarity make it more concrete. In this work, experienced people's interaction is done, where 90% said that splitting is required, and 10% said they have no idea about splitting. The challenge is faced when splitting is going to be done, which involves dependencies between stories.

Raza [17] discussed Scrum and other Agile methods that have become super important in the business scene, helping teams adapt quickly. The proposed work breaks down complex projects into smaller, manageable chunks, turning everything into a handful of user stories that are estimated in "story points." However, predicting these points can be a bit tricky and often leads to inefficiencies.

Wautelet et al. [18] suggested that user stories are a key part of Agile software development. The work captures requirements from the user's viewpoint in a straightforward, user-friendly way. Typically, follow a template that covers the "who," "what," and sometimes "why," keeping developers focused on user needs and expectations. This method encourages ongoing feedback and helps teams adapt quickly during development.

Lucassen et al. [1] worked on the quality of user stories as per the requirement engineering of Agile. The proposed work analyzed the 1023 user stories which are taken from 18 case study organizations. This paper explained why user stories are needed to be further split to make the development easy for Agile developers. The splitting of the user story can be done on the basis of conceptual, estimation, and role.

Yang et al. [19] explained that user stories are super important in Agile development because they help teams stay focused on what customers really need. The suggested work breaks things down into a simple "who, what, why" format, which is great for storytelling. But as projects expand, the number of user stories can get pretty overwhelming, so it's crucial to organize them for effective planning and development. Right now, the way we cluster user stories can be pretty tedious and relies a lot on personal judgment. Thus, it becomes important to automate the clustering of similar user stories using natural language processing.

Nasiri et al. [20] investigated that user stories are super straightforward but really impactful when it comes to outlining what users need from software in Agile development. The work is framed from the user's viewpoint in a simple way: "As a project manager, I want to track the team's progress so that I can ensure we meet our deadlines." This approach keeps the development team laser-focused on what truly matters – the user experience and their objectives. Unlike a rigid requirements document that stays the same throughout the project, user stories are adaptable and can change as needed. Because clearer and more aligned with real-world usage, user stories help everyone – from developers to stakeholders – grasp how to create something that genuinely meets users' needs. Ultimately, well-crafted user stories help keep everyone on the same page and lead to software that's both valuable and satisfying.

Taibi et al. [21] proposed a comparison between the decomposition of the requirements in Agile scrum, XP, Kanban, scrum with Kanban, and Banana development process. This work describes that splitting can be done on the basis of datatypes and parameters. Recommendation of user story splitting for browser compatibility to allow the product owner to prioritize work, as explained in this work.

Santos et al. [22] contributed equally to this work. A well-crafted user story should kick off with some context, dive into the user's needs, and wrap up with a clear takeaway on its value. It usually starts by identifying who the user is. For instance, "As a warehouse manager, I want to track my inventory in real-time so I can prevent shortages." This format is key in Agile development because it highlights the user's goal and its broader impact on operations or efficiency.

Shahid [23] proposed a novel approach for machine learning based user story splitting. The considered ML techniques are random forest, SVM, KNN, and decision tree. To balance the dataset, the use of SMOTE and ROSE techniques is necessary so that they make better predictions. This paper focused on splitting the user story, which is a challenging task, whether splitting is required or not. Because a badly written user story is not easy to implement, it can lead to the delay of completion. That's why splitting is required to complete the task on time.

Peña et al. [24], in Agile development, user stories play a crucial role in helping teams capture key user needs within issue management systems like Jira. However, inconsistent labelling and lack of oversight often lead to many stories being misclassified and lost among other issues, like bug fixes or feature requests. Hence, it becomes important to resolve this issue and automatically identify and categorize user stories from large and varied datasets.

Hudda et al. [25] user stories play a crucial role in the Agile development process. The work helped clarify what users need and why it matters. For example, consider this story: "As a teacher, I want to see my students' attendance so I can monitor their regularity in class." It clearly outlines the user's desire and the reasoning behind it. However, figuring out which story to tackle first can be quite challenging. While understanding the client's top preferences is essential, it's equally important to consider the effort and complexity involved in development. The goal is to prioritize high-impact features that are less complex to implement first.

Arachchi et al. [26] introduce a deep learning model that merges long short-term memory (LSTM) and recurrent highway networks (RHN) to forecast story points for Agile projects. It tackles the problem of demanding the effort needed for user stories. The model was trained from start to finish without needing any manual feature engineering and was evaluated using a dataset of 6,801 issues from open-source projects. The findings showed that it was more accurate than older methods, leading to a decrease in mean absolute error (MAE). This improvement boosts the accuracy of Agile estimations, which helps with planning and managing resources in projects.

Abadeer et al.'s [27]research looks into Deep-SE, a deep learning model that merges LSTM with RHNs to enhance story point estimation in Agile development. By analyzing text from user stories, Deep-SE surpassed conventional models, recording a mean absolute error of 1.46 in a healthcare anonymization project. Still, the model struggled with intricate tasks and unclear user stories, indicating that clearer descriptions are necessary before using machine learning. Future efforts will aim to boost the model's interpretability and scalability for various projects.

Imhmed et al. [28] outlined what customers or stakeholders are looking for, which helps developers grasp and create the right features. Since Agile often involves shifting requirements, manually writing test steps can be a bit of a drag and not very efficient. It can automate tasks like creating test cases or identifying duplicate stories, which saves time and streamlines the whole process. NLP allows teams to move quicker, reduces the time spent on managing user stories, and ensures those stories are clear and effective. Research shows that a whopping 90% of people believe good communication between teams and stakeholders boosts product quality, and 79% think that automating the testing of user stories benefits developers. This highlights how leveraging NLP can help Agile teams consistently deliver top-notch software and keep customers satisfied.

Nasiri et al. [29] introduced a fresh approach to creating business process models from user stories in Agile development. It leverages NLP to pull out essential components such as activities, actions, and conditions from user stories. By applying Prolog-based rules via Python, it generates UML activity diagrams. This technique not only achieves high accuracy in model creation but also shifts the focus from conventional areas like test cases to more intricate business processes. Looking ahead, there are plans to enhance the detection of conditions and synonyms, ultimately aiming to incorporate automated code generation for business services.

Lucassen et al. [30] presented the quality user story (QUS) framework, which outlines 14 essential quality criteria for assessing user stories. These criteria are designed to enhance the quality and clarity of user stories in Agile development. Additionally, the authors created AQUSA, an automated tool that leverages the QUS framework to spot issues in user stories, assisting teams in boosting their quality. In early tests, AQUSA demonstrated impressive accuracy in identifying problems like non-atomic and poorly structured stories. The study indicates that AQUSA could play a crucial role in improving user stories, with future efforts aimed at broadening its capabilities, reducing false positives, and assessing its long-term effectiveness in real-world projects.

Gilson et al. [31] looked into how NLP can be used to turn user stories into UML use case diagrams automatically, which helps make requirements clearer and easier to validate in Agile development. By visualizing these user stories, teams can spot problems more easily and boost communication. The goal is to minimize mistakes from unclear descriptions and make the validation process smoother. Future efforts will focus on assessing how accurate and effective these automated diagrams are in actual projects.

Singla et al. [32] looked into how Scrum can be tailored for machine learning (ML) and data science initiatives, tackling issues such as erratic workflows and the necessity for ongoing experimentation. The team examined Jira logs and employed an SVM model to categorize user stories as either "good" or "bad," which enhanced the quality of the stories and the logging of tasks. It indicates that ML projects gain from more adaptable Scrum methods and suggests modifying tools like Jira to better accommodate the fluid nature of ML tasks. Future studies will evaluate how these changes affect team productivity and project results over time, ensuring Scrum remains effective for various project types.

The above discussion states that various gaps exist to explore the work of splitting user stories in ASD.

## 5.4 Techniques to Split User Stories

Larger user stories are split into smaller ones by using various techniques. Breaking down large user stories into manageable user stories; the vertical and horizontal slicing are two approaches [33] [34]. To work with smaller parts, then use horizontal slicing, and to work with the delivery of end-to-end functionality, then proceed with vertical slicing. To deliver value quickly, reducing the risk, collaboration enhanced, improved visibility, these are the

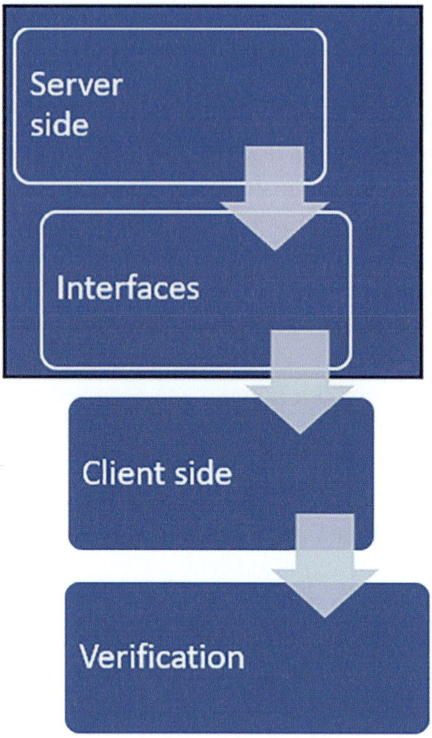


**Figure 5.1** Horizontal slicing.

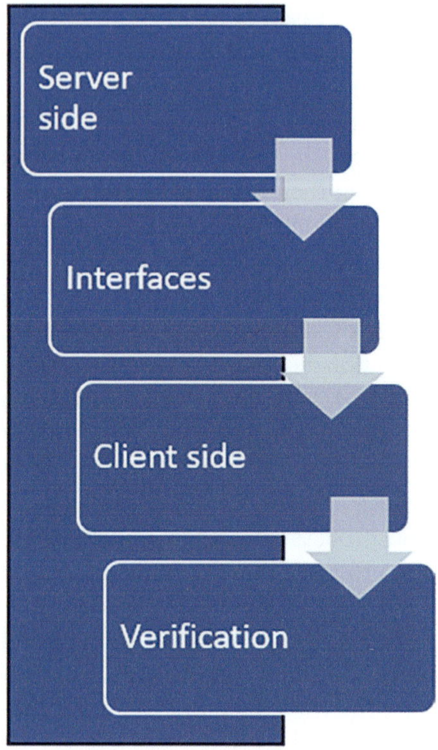


**Figure 5.2** Vertical slicing.

benefits the Agile team can give in ASD [35, 36]. Figures 5.1 and 5.2 showed the different layers of project as server-side, prepare interfaces, development of client side, and verification where horizontal and vertical slicing represented their work.

For example, user wants to login then needs server-side, interfaces and client-side to be developed, collaborate interface with client-side, and proceed with verification. For horizontal slicing, the following user stories need to write for server-side, prepare interfaces, develop the client-side, and finally verify. For vertical slicing, the following stories are to be broken down so that they must be able to login with valid/invalid credentials and password conditions [37].

The following techniques [38] are used to split the user stories.

- User roles and personas
- Workflow

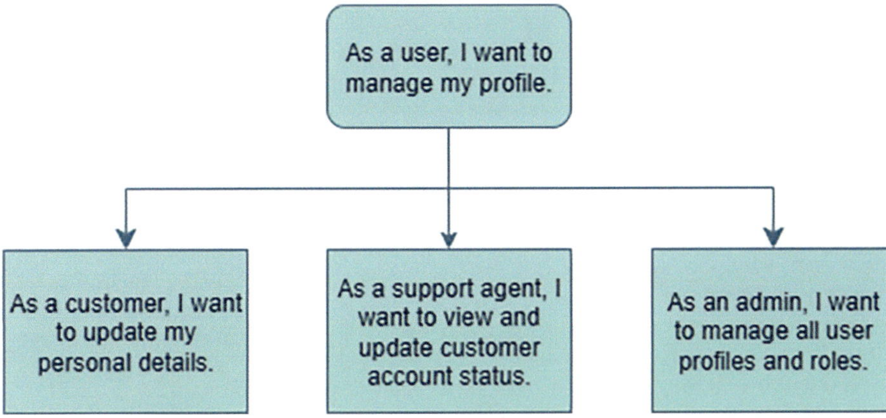


**Figure 5.3** User roles and personas.

- Platforms
- Data variation
- Operations (CRUD)
- Acceptance criteria
- Data types
- Business rules

### 5.4.1 User roles and personas

In this technique, the role specified in user stories can take a step to be further split. For example, a user can manage the profiles, where roles can be split into customer, support agent, and admin, as shown in Figure 5.3. When I want to update the personal details, view the customer status, and manage the profiles and roles.

### 5.4.2 Workflow

The flow of work can also be broken down into steps to complete the task. When a user requirement explains the working step from initial to final, then it should be managed as per the given task. For example, Figure 5.4 shows that the user wants to submit the feedback, it has a flow from submission to receive and view the submitted feedback. They also need to split on behalf of their working flow from first step to final step.

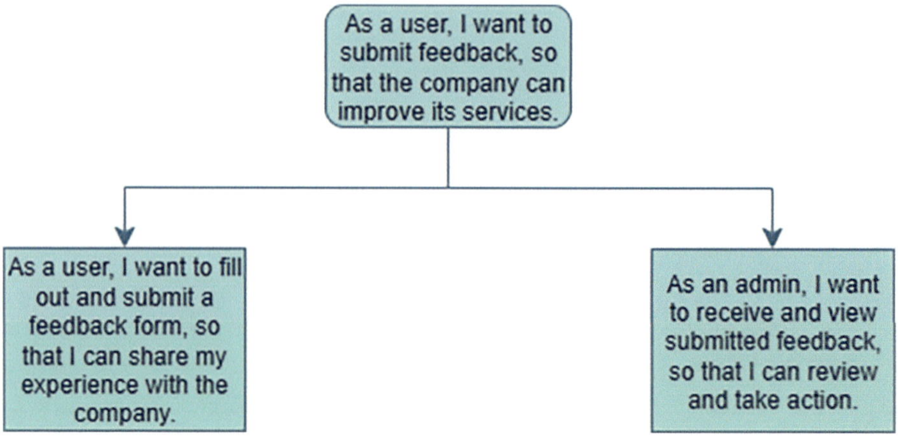


**Figure 5.4** Workflow.

### 5.4.3 Business rules

Business rules are the key principles that define the rules and regulations for the users. A user can apply for a loan to get financial support, but as a user first applies for the loan, then gets a notification about approval or rejection of the loan application based on the rules of business. So, business rules can also split the user story as per their access rights, splitting as shown in Figure 5.5.

### 5.4.4 Based on data variations

How to enter the data into software based on that user story, which also splits into their defined variation of data. If, as a customer or user, I want to view

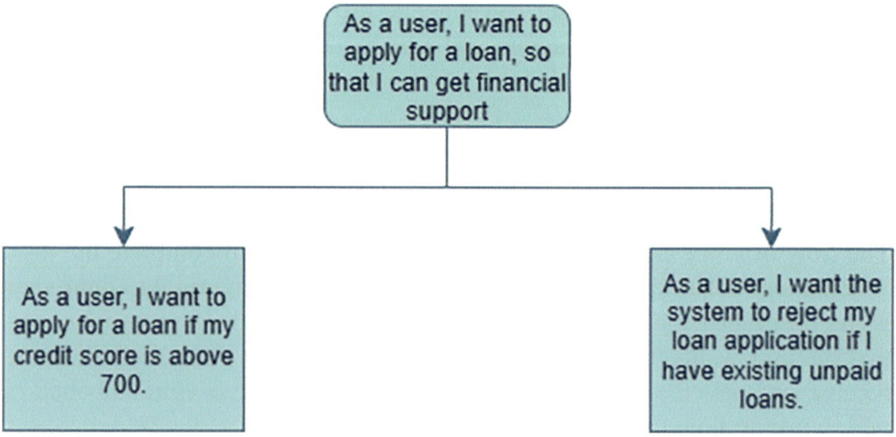


**Figure 5.5** Business rules.

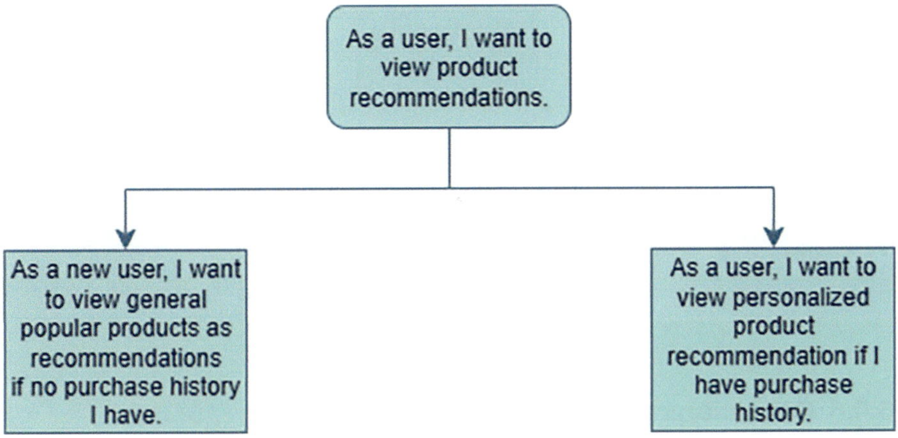


**Figure 5.6** Based on data variations.

the recommendations of products, then it should be dependent on the user's purchase history, as shown in Figure 5.6.

### 5.4.5 Platform

A software can be developed for different platforms (differ across platforms) as per their requirement. Worldwide, the application can be accessed by different platforms, such as through web user, mobile application user, and desktop application user, as represented in Figure 5.7. Platform-based splitting can also be done to make easy development of Agile software to support platform independence.

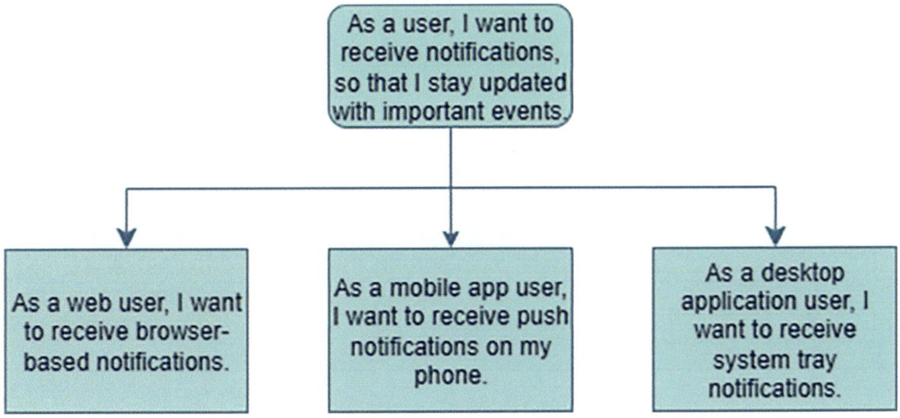


**Figure 5.7** Platform.

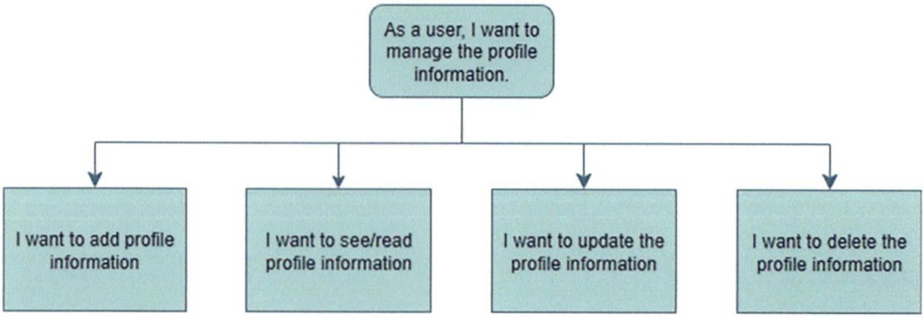


**Figure 5.8** Operations (CRUD).

### 5.4.6 Operations (CRUD)

Distinguish operational query also performed to manage the data in software. These operations are Create, Read, Update, and Delete to maintain the data into the database. Each operation has their story point by estimation to complete the user story on time [39]. It is mostly used when the user story has interaction with the backing store or backend, which is depicted in Figure 5.8.

### 5.4.7 Acceptance criteria

Spitting is done on behalf of conditional statements where conditions are met. Each acceptance criterion represents a testable requirement that must be satisfied before the user story can be delivered to stakeholders. Figure 5.9 shows how to reset the password.

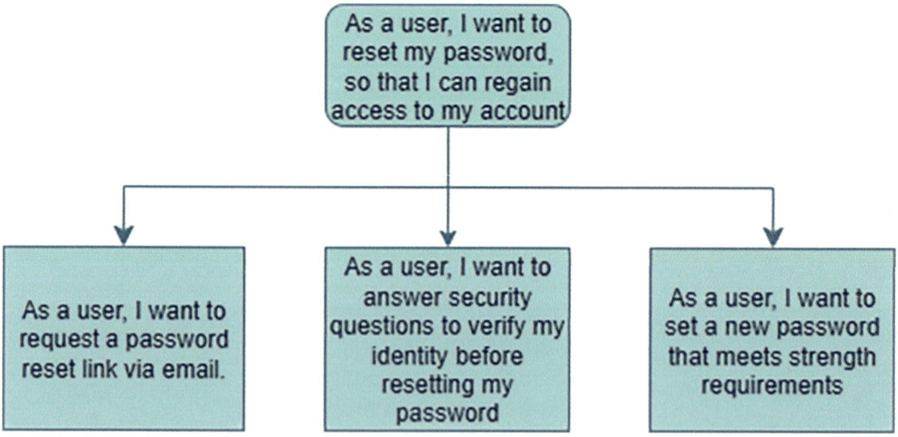


**Figure 5.9** Acceptance criteria.

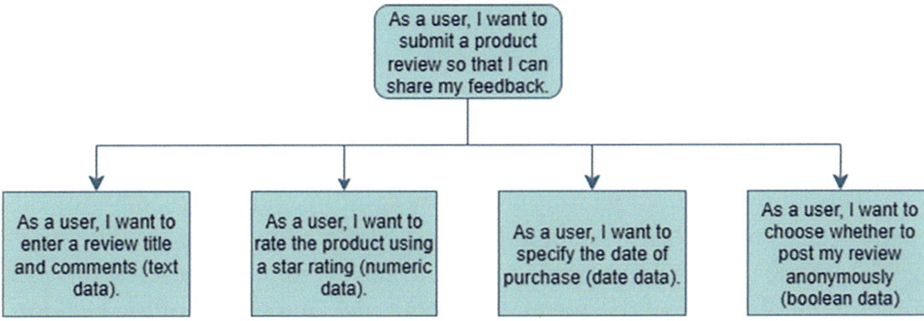


**Figure 5.10** Datatypes.

### 5.4.8 Datatypes

Different type-based splitting can also happen when it is required, as per the customer satisfaction or application requirement. For example, a user wants to submit the review of a product that can be in textual format, rate as numeric, date of shopping, and some Boolean data also. Later, it has been broken down by their data types for the convenience of the application shown in Figure 5.10.

Figure 5.11 shows that splitting techniques are used in this work, where each type is explained with suitable examples. User roles, platform,

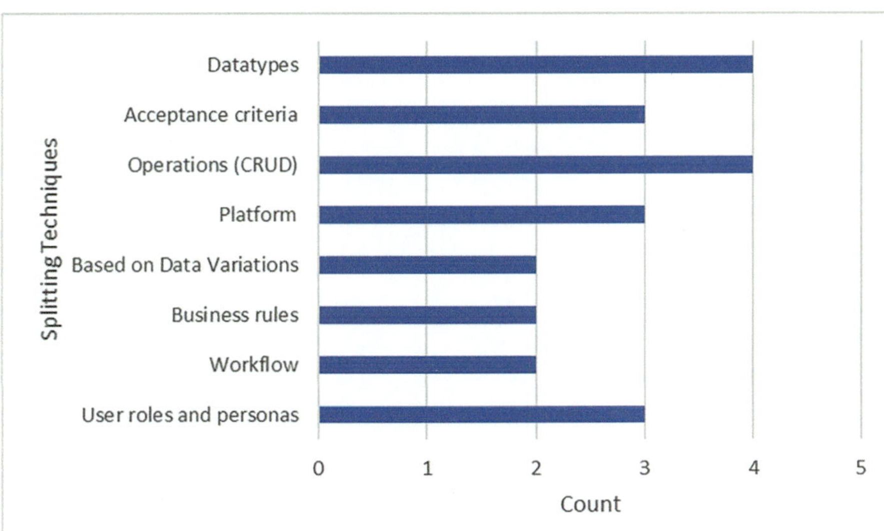


**Figure 5.11** User stories splitting techniques (user story count).

and acceptance criteria techniques split the user story into three user stories, whereas workflow, business rule, and data variation techniques split the user story into two, and finally operations and datatypes both split into four user stories. These techniques definitely increase the number of user stories but improve the ease of the development process of Agile.

This study explored various user story splitting techniques, demonstrating how large user stories can be effectively broken down into smaller, manageable stories using appropriate and relevant examples. These techniques improve the planning of the sprint, team outcome, and clarity of the backlog.

## 5.5 User Story Splitting Impact on Agile Software Development Process

Positive impacts of user story splitting are explained in Figure 5.12. The positive points are faster feedback delivery, prediction improvement, risk reduction, collaboration, and focus are also improved; prioritization is also flexible [40]. Manageable stories and progress increased, accuracy improved for estimation, issues are identified earlier to make testing, Agile development team.

Collaboration worked well, and prioritization is also done at an early stage of development [41]. Negative impacts of user story splitting are

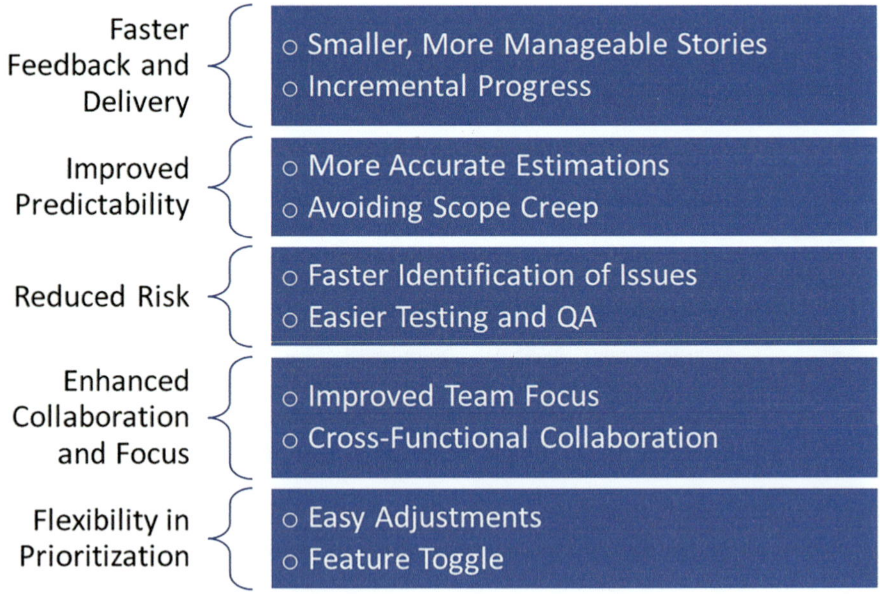


**Figure 5.12** Positivity of user story splitting.

explained in Figure 5.13. Managing user stories is also complex; overhead increases the work, challenges in integration of work, risk of duplicity, and dependencies to manage are also complex for splitting.

Research directions for over-splitting are recommended to investigate the adaptive framework that can split the user story in an optimal way. Splitting of the user story is balancing the necessary details and also avoiding the over-splitting. Another scope is to work on context switching because it causes the loss of focus, increases the cognitive load, which leads to disrupting the flow and reduces the sprint efficiency. Cohesion-based splitting model that ensures the integrity of functionality, which can improve the delivery quality.

Overall discussion of user stories splitting methods is commonly used to achieve efficient use of the Agile development team. The splitting takes place when the complex user story leads to a degradation of the performance of the team. Splitting is also done on behalf of various factors, such as operational, workflow, acceptance criteria, variation, and platform type of requirements involvement in ASD. Smaller user stories can be completed on time, but the larger user story takes time to complete, and may require overtime completion. So, splitting should be done as early as possible from the perspective

**Figure 5.13** Negativity of user story splitting

of developers. All the pros and cons are also discussed when the splitting is considered. Splitting accelerates the development process of Agile software, which leads to improvement in customer satisfaction. Splitting is also giving the optimized solution to complete the user story on time, so that Agile software handover to the client is on time.

## 5.6 Example

Assign the metrics to each sub story with the following parameters: completion rate, sprint velocity, defects, and cycle time [42, 43]. The completion rate is settled by assigning the percentage rate, such as for a large user story 70–80%, 85–95% for medium, and for a small user story, assign 95–100% a base completion probability. Sprint velocity depends on the average story points delivered per sprint for a user story, which is the sum of completed story points per sprint. For example, 8 story points are given to the original user story, if split into 4, then 2 for each sub-user story. If ¾ completed, then velocity is equal to 6 story points, which says velocity = 6/one sprint = 6. Now the third parameter defect is reported per user story, where for large user stories the defect value is 1.0–1.5, and for small the value is 0.5 to 1.0. Last but not the least parameter is cycle time, which is represented by the average time taken to complete a user story or sub-user story. For large user stories 5–7 days, for medium 3–5 days, and 1–3 days for smaller user stories. For example, if a large story cycle time is 6, then split into 3, where each sub-user story takes 2–3 days, then the average cycle time is 2.5 days.

For example, user story is taken to compute the above parameters US1: As a user, I want to reset my password, so that I can recover my account. This user story comes under the workflow splitting type. If this is the US1, which involves the reset request, send reset email, set new password that is why split count is 3 which is also come under medium that shows the 95% completion probability, say to reset request 3 story point, email 2 story point, and set new password the story point is 3 then the total points are 8. If sprint-wise story points are completed 8 points in a single sprint, then next may be 7 points completed in the next sprint, next time again 8, and then 10 may be taken to complete the sprint. The average velocity = (8 + 7 + 8 + 10) ÷ 4 + 33 ÷ 4 = 8.25, approximately 8.2. The defect value is 0.8 as per below. As for medium-sized user stories, the average cycle time is 3.5 days, which is between 3 and 5 days.

This example shows how the splitting is optimized in the process of Agile software development, which improved the sprint velocity, increased the completion rate, and also reduced the cycle time for each large-size user story.

## 5.7 Conclusion

In conclusion, splitting of user story is a powerful tool in ASD, especially in Scrum. Splitting focused on team productivity, work quality, and customer satisfaction. Splitting sliced the user story vertically and horizontally; it depends on the application type. Splitting has its positive and negative impacts on the ASD process. This work discussed the use of splitting may improve the Agile team performance as velocity of sprint. Each iteration is completed on time when the splitting of user stories takes place at the right moment. Splitting is done when developers face the challenges of implementation of complex user stories due to time rigidness. On-time completion of requirements is the prime goal of Agile software. So, need to complete the task on time instead of late completion. Otherwise, the late work delays the overall development time of the software. This work shows the comparison between splitting techniques. Some of the splitting techniques split the user story into 2, 3, and 4-number user stories to make the development process. By adopting a structured approach to splitting user stories, teams can ensure smoother project execution, better stakeholder satisfaction, and a more sustainable pace of work over time. In other ways the velocity of sprint is improved, the completion rate is also increased, and the cycle time of the user story is reduced. It is recommended to develop adaptive frameworks for optimal user story splitting, address context-switching issues to reduce cognitive load, and design cohesion-based models to preserve functionality and improve sprint outcomes.

## References

[1] G. Lucassen, F. Dalpiaz, J. M. E. M. van der Werf, and S. Brinkkemper, "Improving agile requirements: the Quality User Story framework and tool," Requir Eng, vol. 21, no. 3, pp. 383–403, 2016, doi: 10.1007/s00766-016-0250-x.

[2] V. KANNAN and D. WILLETT, "User stories as lightweight requirements for agile clinical decision support development," Agile Alliance, vol. 26, no. 11, pp. 1344–1354, 2019, doi: 10.1093/jamia/ocz123.

[3] M. A. Kuhail and S. Lauesen, "User Story Quality in Practice: A Case Study," Software, vol. 1, no. 3, pp. 223–243, Jun. 2022, doi: 10.3390/software1030010.

[4] B. Seitlheko and L. Mwansa, "Exploring the Decomposition of Epics using Natural Language Processing," INFORMATION COMMUNICATION TECHNOLOGY (ICT) JOURNAL, vol. 6, no. 1, pp. 1–6, 2022.

[5] S. Alsaqqa, S. Sawalha, and H. Abdel-Nabi, "Agile Software Development: Methodologies and Trends," International Journal of Interactive Mobile Technologies (iJIM), vol. 14, no. 11, p. 246, Jul. 2020, doi: 10.3991/ijim.v14i11.13269.
[6] S. Dimitrijević, J. Jovanović, and V. Devedžić, "A comparative study of software tools for user story management," Inf Softw Technol, vol. 57, pp. 352–368, Jan. 2015, doi: 10.1016/j.infsof.2014.05.012.
[7] P. Weichbroth, "A Case Study on Implementing Agile Techniques and Practices: Rationale, Benefits, Barriers and Business Implications for Hardware Development," Applied Sciences, vol. 12, no. 17, p. 8457, Aug. 2022, doi: 10.3390/app12178457.
[8] B. Kumar, U. Tiwari and D. C. Dobhal, "User Story Splitting in Agile Software Development using Machine Learning Approach," *2022 Seventh International Conference on Parallel, Distributed and Grid Computing (PDGC)*, Solan, Himachal Pradesh, India, 2022, pp. 167-171, doi: 10.1109/PDGC56933.2022.10053226.
[9] C. Verwijs and D. Russo, "A Theory of Scrum Team Effectiveness," ACM Transactions on Software Engineering and Methodology, vol. 32, no. 3, pp. 1–51, Jul. 2023, doi: 10.1145/3571849.
[10] M. L. Drury-Grogan, "Performance on agile teams: Relating iteration objectives and critical decisions to project management success factors," Inf Softw Technol, vol. 56, no. 5, pp. 506–515, May 2014, doi: 10.1016/j.infsof.2013.11.003.
[11] K. Dikert, M. Paasivaara, and C. Lassenius, "Challenges and success factors for large-scale agile transformations: A systematic literature review," Journal of Systems and Software, vol. 119, pp. 87–108, Sep. 2016, doi: 10.1016/j.jss.2016.06.013.
[12] H. Saeeda, M. O. Ahmad, and T. Gustavsson, "Challenges in Large-Scale Agile Software Development Projects," in Proceedings of the 38th ACM/SIGAPP Symposium on Applied Computing, New York, NY, USA: ACM, Mar. 2023, pp. 1030–1037. doi: 10.1145/3555776.3577662.
[13] M. Adzgauskaite, C. Tam, and R. Martins, "What helps Agile remote teams to be successful in developing software? Empirical evidence," Inf Softw Technol, vol. 177, p. 107593, Jan. 2025, doi: 10.1016/j.infsof.2024.107593.
[14] D. Ghimire and S. Charters, "The Impact of Agile Development Practices on Project Outcomes," Software, vol. 1, no. 3, pp. 265–275, Aug. 2022, doi: 10.3390/software1030012.
[15] P. Pokharel and P. Vaidya, "A Study of User Story in Practice," in 2020 International Conference on Data Analytics for Business and Industry:

Way Towards a Sustainable Economy (ICDABI), IEEE, Oct. 2020, pp. 1–5. doi: 10.1109/ICDABI51230.2020.9325670.

[16] O. Liskin, R. Pham, S. Kiesling, and K. Schneider, "Why we need a granularity concept for user stories," in Lecture Notes in Business Information Processing, Springer Verlag, 2014, pp. 110–125. doi: 10.1007/978-3-319-06862-6_8.

[17] A. Raza, "Predicting the Duration of User Stories Machine learning for agile planning," Arcada University of Applied Sciences, 2020.

[18] Y. Wautelet, S. Heng, S. Kiv, and M. Kolp, "User-story driven development of multi-agent systems: A process fragment for agile methods," Comput Lang Syst Struct, vol. 50, pp. 159–176, Dec. 2017, doi: 10.1016/j.cl.2017.06.007.

[19] B. Yang, X. Ma, C. Wang, H. Guo, H. Liu, and Z. Jin, "User story clustering in agile development: a framework and an empirical study," Front Comput Sci, vol. 17, no. 6, Dec. 2023, doi: 10.1007/s11704-022-8262-9.

[20] S. Nasiri, Y. Rhazali, M. Lahmer, and N. Chenfour, "Towards a Generation of Class Diagram from User Stories in Agile Methods," Procedia Comput Sci, vol. 170, pp. 831–837, 2020, doi: 10.1016/j.procs.2020.03.148.

[21] D. Taibi, V. Lenarduzzi, A. Janes, K. Liukkunen, and M. O. Ahmad, "Comparing requirements decomposition within the Scrum, Scrum with Kanban, XP, and Banana development processes," in Lecture Notes in Business Information Processing, Springer Verlag, 2017, pp. 68–83. doi: 10.1007/978-3-319-57633-6_5.

[22] C. A. dos Santos, K. Bouchard, and B. Minetto Napoleão, "Automatic user story generation: a comprehensive systematic literature review," Int J Data Sci Anal, Jun. 2024, doi: 10.1007/s41060-024-00567-0.

[23] M. B. Shahid, "Splitting User Stories Using Supervised Machine Learning," UNIVERSITY OF TARTU, 2020.

[24] F. J. Peña, M. L. Roldán, and M. M. Vegetti, "Identification of user stories in software issues records applying pre-trained natural language processing models," in Congreso Nacional de Ingeniería Informática / Sistemas de Información (CONAIISI 2020), 8o CONAIISI, 2020.

[25] S. Hudda, R. Mahajan, and S. Chopra, "Prioritization of User-Stories in Agile Environment," Indian J Sci Technol, vol. 9, no. 45, Dec. 2016, doi: 10.17485/ijst/2016/v9i45/105069.

[26] K. Wanigasooriya Arachchi et al., "An agile project management supporting approach for estimating story points in user storie," in 8th International Conference in Information Technology Research 2023

Moratuwa, Sri Lanka, Information Technology Research Unit, Faculty of Information Technology, University of Moratuwa., 2023.

[27] M. Abadeer and M. Sabetzadeh, "Machine Learning-based Estimation of Story Points in Agile Development: Industrial Experience and Lessons Learned," in 2021 IEEE 29th International Requirements Engineering Conference Workshops (REW), IEEE, Sep. 2021, pp. 106–115. doi: 10.1109/REW53955.2021.00022.

[28] H. Imhmed, K. Ahmed, Y. Salem, and H. Zulzalil, "Leveraging Latent Natural Language Processing Techniques for User Story Management in Agile Software Development," Journal of Pure & Applied Sciences, vol. 22, no. 2, pp. 5–9, Oct. 2023, doi: 10.51984/jopas.v22i2.2599.

[29] S. Nasiri, A. Adadi, and M. Lahmer, "Automatic generation of business process models from user stories," International Journal of Electrical and Computer Engineering (IJECE), vol. 13, no. 1, p. 809, Feb. 2023, doi: 10.11591/ijece.v13i1.pp809-822.

[30] G. Lucassen, F. Dalpiaz, J. M. E. M. van der Werf, and S. Brinkkemper, "Forging high-quality User Stories: Towards a discipline for Agile Requirements," in 2015 IEEE 23rd International Requirements Engineering Conference (RE), IEEE, Aug. 2015, pp. 126–135. doi: 10.1109/RE.2015.7320415.

[31] F. Gilson and C. Irwin, "From User Stories to Use Case Scenarios towards a Generative Approach," in 2018 25th Australasian Software Engineering Conference (ASWEC), IEEE, Nov. 2018, pp. 61–65. doi: 10.1109/ASWEC.2018.00016.

[32] K. Singla, T. M. Vinayak, A. S. Arpitha, C. Naik, and J. Bose, "Story and Task Issue Analysis for Agile Machine Learning Projects," in 2020 IEEE-HYDCON, IEEE, Sep. 2020, pp. 1–4. doi: 10.1109/HYDCON48903.2020.9242803.

[33] I. M. Ratner and J. Harvey, "Vertical Slicing: Smaller is Better," in 2011 AGILE Conference, IEEE, Aug. 2011, pp. 240–245. doi: 10.1109/AGILE.2011.46.

[34] O. Dieste et al., "The role of slicing in test-driven development," Jul. 2024, Accessed: Jan. 24, 2025. [Online]. Available: 10.48550/arXiv.2407.13258

[35] A. S. Ghai, D. Kapil, L. Sejwal, A. Gupta, and S. Devliyal, "Use of agile software development method to manage academic resources," in AIP Conference Proceedings, 2022, p. 020021. doi: 10.1063/5.0103745.

[36] L. Barros, C. Tam, and J. Varajão, "Agile software development projects–Unveiling the human-related critical success factors," Inf Softw Technol, vol. 170, p. 107432, Jun. 2024, doi: 10.1016/j.infsof.2024.107432.

[37] Sujith G, "What Is User Stories In Agile? And Techniques To Split Them," https://benzne.com/blogs/agile/user-stories-techniques/.

[38] E. Dellsén, K. Westgårdh, and J. Horkoff, "Invest in Splitting: User Story Splitting Within the Software Industry," in Requirements Engineering: Foundation for Software Quality. REFSQ 2022, 2022, pp. 115–130. doi: 10.1007/978-3-030-98464-9_10.

[39] M. Choetkiertikul, H. K. Dam, T. Tran, T. Pham, A. Ghose, and T. Menzies, "A Deep Learning Model for Estimating Story Points," IEEE Transactions on Software Engineering, vol. 45, no. 7, pp. 637–656, Jul. 2019, doi: 10.1109/TSE.2018.2792473.

[40] S. I Kennedyd, A. A. Zadeh, J. Choi, and S. Alborz, "Agile Practices and IT Development Team Well-Being: Unveiling the Path to Successful Project Delivery," Engineering Management Journal, pp. 1–13, Oct. 2024, doi: 10.1080/10429247.2024.2413710.

[41] D. Batra, W. Xia, and M. Mingyu ", "Collaboration in Agile Software Development: Concept and Dimensions," Communications of the Association for Information Systems, vol. 41, pp. 429–449, 2017, doi: 10.17705/1CAIS.04120.

[42] E. N. BUDACU, P. POCATILU, "Real Time Agile Metrics for Measuring Team Performance", Informatica Economica 22(4/2018):70-79, DOI: 10.12948/issn14531305/22.4.2018.06.

[43] R. Yegyanarayanan, "Building Agile Metrics Right: A Guide to Effective Measurement in Agile Teams - Measure What matters", Conference Proceeding Issue Published in International Journal of Trend in Research and Development (IJTRD), ISSN: 2394-9333, www.ijtrd.com, Agile Research Conference (ARC-24), October, 2024.

# 6

# Enhancing Data Security through Secure Data Transactions: Techniques and Future Directions

**Archana Kero[1] and Suman Pant[2]**

[1]Department of Computer Science, Sai Institute of Paramedical and Allied Sciences, India
[2]School of Science and Technology, Swami Rama Himalayan University, India

**Email:** archanakero@gmail.com; sumanpant@srhu.edu.in

## Abstract

A username and a secret phrase combination is used by many banks and other services to verify customers. Confirmation is a fundamental component of obtaining any online banking system. It becomes a tedious and ineffective chore to keep track of login credentials for several records. Furthermore, there is a chance that users, institutions, or confirmation servers could conduct a range of attacks against traditional verification systems, which have a history of failure. Long-term information leak reports highlight how attackers have developed several innovative ways to obtain clients' certifications, which might pose a real risk. The current article presents a practical and efficient client verification scheme that leverages standalone devices that leverage several Rivest-Shamir-Adleman cryptographic features, such as hashing, encryption, and digital signatures. The technique takes advantage of various intelligent wearable and portable devices as well as ubiquitous processing to enable users to carry out a secure verification convention. Utilizing secret word tables and static usernames to identify and validate the legitimacy of login clients, our suggested scheme eliminates the requirement for a verification waiter. It can fend off replay attacks, shoulder-riding attacks, phishing



DOI: 10.1201/9788743812739-6

attacks, and information breach incidents in addition to being safe from attacks involving secret keys.

## 6.1 Introduction

### 6.1.1 Background and motivation

The use of usernames and passwords in traditional authentication schemes puts online financial services, financial systems, and their clients at serious risk. Most contemporary verification systems either assign or let a client choose a distinct, static client ID that functions as a mark. The client has frequently been connected to this static name for a while. Unexpectedly, consumers tend to utilize the same client ID on many websites and frameworks. Moreover, a lot of individuals still use the same password for a variety of databases and online accounts. A recent poll found that 51% of the clients analyzed reuse similar secret keys across many websites, and over 77% of members either slightly tweak or reuse their present passwords [1]. Figure 6.1 shows the distinct common attack vectors in the field of data transactions.

### 6.1.2 Security challenges with traditional systems

Security risks, such as insider attacks, could result from this common practice. Insiders and nefarious individuals can access many websites and

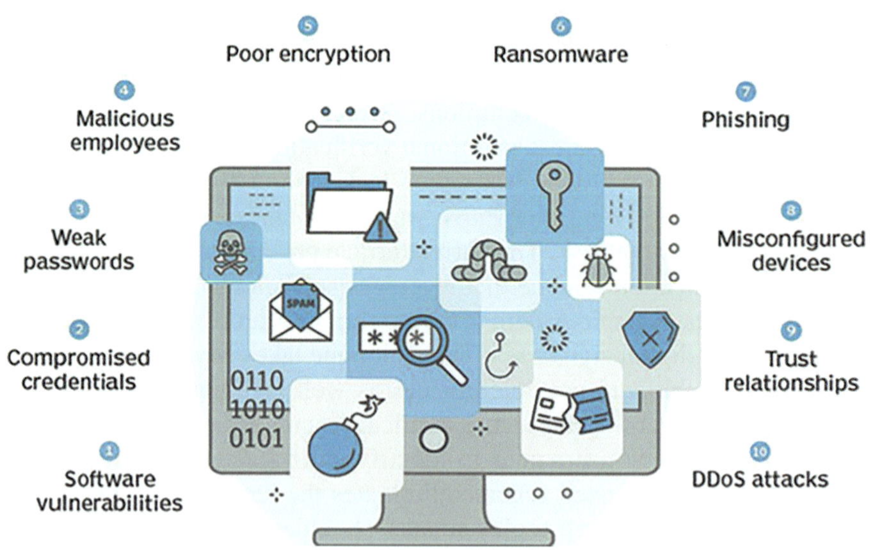


**Figure 6.1** Common attack vector.

administrations by using usernames and password-protected word tables. Bitter insiders might even make money by reselling this sensitive information on the dark web for bitcoin or other untraceable cryptocurrency payments through schemes like Zero coin or Bitcoin [2]. A phisher may also be able to use their clients' credentials on several websites with this training. Phishing is a type of friendly design attack in which a malicious customer, often known as a phisher, attempts to masquerade as a trustworthy entity or public organization to fraudulently obtain the certifications of legitimate clients [3, 4]. Phishing attacks usually use special correspondence methods, such as email or SMS, and their goal is to trick the victim into visiting a fraudulent website that looks exactly like the legitimate one.

To deal with the security risks during data transactions, a continuous process is shown in Figure 6.2. The figure represents a risk management process, which is typically structured in a cyclic manner to ensure continuous improvement and adaptation. The modules of the risk management can be further segregated into different modules, such as:

- Context Establishment: This is the starting point of the risk management process. It involves defining the external and internal context within which the risk management process will be implemented. This includes understanding the organizational environment, stakeholders, risk criteria, and the scope of the risk assessment.
- Risk Assessment: This central module comprises three steps:
    - Risk Identification: In this step, all potential risks that could affect the objectives are identified. This involves recognizing and documenting risks.
    - Risk Analysis: After identification, the risks are analyzed to understand their nature, causes, and potential consequences. This helps in understanding the level of risk.
    - Risk Evaluation: Based on the analysis, risks are evaluated to determine their significance and decide which risks need to be treated and in what order of priority.
- Risk Treatment: After evaluating the risks, this module involves selecting and implementing measures to mitigate, transfer, avoid, or accept the risks. The goal is to bring the risk level within the organization's risk appetite.
- Communication and Consultation: Throughout the process, communication and consultation with stakeholders are essential. This ensures

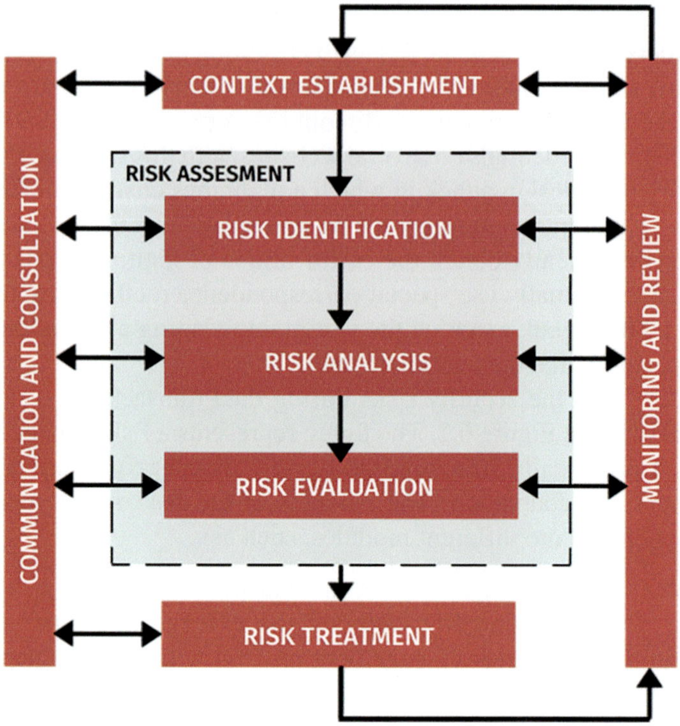


**Figure 6.2** Risk management during data transaction [5].

that risk information is shared effectively, and stakeholders' perspectives are considered in decision-making.

- Monitoring and Review: Risk management is an ongoing process, and this module emphasizes the importance of continuously monitoring the risks and the effectiveness of the risk treatment strategies. Regular reviews ensure that the risk management process remains relevant and effective over time.

### 6.1.3 The need for enhanced security measure

In a phishing attack, a fraudster may choose to target a single client or a group of clients, obtain their login credentials, and then attempt to access basic services like internet banking. Static accreditations are a major issue that contributes to the success of phishing attacks [6]. It may be possible to alter the game's rules and generate more potent anti-phishing verification techniques by adopting a different viewpoint and giving up on using static usernames and passwords [7].

By recommending a one-time username confirmation along with a secure check code for every login session, we demonstrate in this work how smart personal devices may improve security and user experience [8]. No need to study complicated passwords or keep track of multiple usernames for the customer. Below is a breakdown of the main responsibilities of this work:

- Rather than clients using their usernames and passwords, we create and implement a unique plan that involves encryption and marking. With respect to inheritance verification procedures, this approach offers a higher level of safety and reduces vulnerabilities.
- We introduce the concept of client-driven admission control, which can play a significant role in security verification and upgrades. Clients have power over client-driven admission control, since they can record their consent for each login meeting.
- We examine the validity of the suggested validation plan and demonstrate its viability and productivity. We specifically investigate the security of the given confirmation scheme from several angles, including phishing attacks, attacks involving secret keys, shoulder-riding attacks, replay attacks, and so forth.
- We demonstrate how our strategy extends validation security by adhering to the one time pad (OTP) property for the confirmation code and meeting key.
- In terms of correspondence and computation overhead, we evaluate how the suggested confirmation conspiracy is presented.

The following is the work schedule, which highlights the most relevant written material as well as our sources of inspiration, explains the plan's goals, the hazard model, and the framework model. We describe the confirmation convention that is being suggested to explain how secure our plan is, as well as the results of tests and the correlation and evaluation of their execution.

Authorization will be done in passive mode. The clients are only able to view the exchanges in the inactive mode; they are unable to engage in any more dynamic activity.

Clients are involved in the establishment of one-time secret phrase permission when consents are indicated, providing a client-driven paradigm for admission control. With the standard of least advantage, it has various advantages, including improved security. Providing a client with only those benefits that are essential to their work is considered in such a plan [9]. Resolving authorizations, however, adds to the workload for clients and necessitates a program for mindfulness preparation.

## 6.2 Literature Review

Online banking services, financial organizations, and their clients are seriously at risk from traditional authentication systems, such as the usage of a login and secret word combination. Thus, a unique strategy was devised that synchronizes encryption and marking without relying on clients to recall usernames and passwords.

The client and the server are the two main components of the present framework model. The client's terminal and the enlisted devices make up the customer side. Gadgets enlisted: A clever personal device, such as a cell phone, that is enrolled is capable of carrying out cryptographic tasks. Before using the server's services, each client must register a device with it. There should be no need for a static login and secret code in order for a valid client to access server services. In the process, we expect that the client has successfully enlisted the server's clever gadget.

The client's terminal: A client's terminal is a computer, workstation, or PC that is used to access the server and examine or perform exchanges. Server: The server is linked to a hardware security module (HSM) that handles cryptography and hides the secret key. It is situated in an area similar to a bank. Along with providing various forms of support, the server distributes to the clients its public key and confirmation code.

With this effort, Alrawais et al. [10] have put forth Fog processing thought to be a highly virtualized viewpoint that can enable computing at the edge of the enterprise – that is, at the internet of things devices – to deliver services and applications more successfully and productively. As haze registration originates from distributed computing and is an inevitable extension of it, it inherits many of the security and protection challenges associated with distributed computing, giving rise to widespread concerns in the research community. In this work, a successful key trade convention based on figure text strategy trait-based encryption (CP-ABE) is used to establish secure exchanges among the members, enabling reliable and confidential correspondences among a group of fog hubs. In this paper, computerized signature techniques with CP-ABE were used to achieve privacy, confirmation, certainty, and access control. The convention's efficacy is analyzed in terms of security and implementation [10].

Bonneau et al. [11] put forward to evaluate 20 years of proposals to exchange text passwords for widely applicable online client authentication using a comprehensive set of 25 features that enhance usability, deployment capability, and security that a well-designed system may provide. Their work covers a wide range of suggestions, such as biometrics, equipment tokens, telephone-supported plans, graphical password plans, integrated login conventions, one-time passwords, equipment tokens, and password board programming. Key insights concerning the challenges of replacing passwords are prompted by comprehensive research. Not only does no executed scheme fall short of providing every perfect benefit, but none even comes close to providing the entire set of benefits that legacy passwords currently provide.

There are various options available, ranging from those that provide no security benefits other than password inheritance to those that give significant security benefits but come with higher costs or are more difficult to use. Because scientists rarely consider a sufficiently broad range of legitimate requirements, therefore, it can be inferred that many academic proposals have failed to gain traction [12].

Due to the lack of smartcard readers on most web devices, Bernd Borchert has advocated in this study that smartcard authentication for online services remains a unique application. Here, we look at a method that lets you log in to nearly any web device using an NFC-enabled smartcard and an NFC-capable smartphone. We explore the security, deployability, and convenience aspects of this method while elucidating its intricacies.

The standard online verification technique as of right now is client and secret phrase validation. Another "what-you-have" feature, such as a smartcard, is appealing for obvious security reasons. The planning of a verification convention for web logins using smartcards is noteworthy; nonetheless, the internet browser's inability to communicate with a peruser, whether present or not, and the absence of smartcard readers restrict the extent of the response. Grosse and Upadhyay suggest giving the application the ability to communicate with a smartcard chip by inserting it into a token with a USB interface or a dual interface (USB and NFC). By keeping the NFC token close by, a customer can utilize this method to sign into their record on his NFCperuser-prepared web customer. This method, which focuses on the NFC network, is referred to as the direct NFC Login strategy. This technique's drawback is that it requires clients to have access to a USB or NFC, which severely restricts the program's expansion. In [13], John Brainard has suggested in this work that three variables are typically used for user confirmation in figure frameworks: something you have (like an equipment token), something you are (like a distinct finger impression), and something you know (like a secret word). The fourth component, the client's informal organization – that is, someone you know – is the subject of our investigation in this study. It is an ingrained practice to obtain human validation from a common associate. It assumes roles in peer-level certification, help desk assistance, renowned organizations, advantage designation, and helpdesk support in the field of PC security. Nonetheless, the reliance of one person on another has little logical basis as a quick way for reasonable confirmation. This paper explores the concept of peer-level human-mediated access control validation, or vouching, and studies its application in crisis verification, where key authenticators such as equipment tokens or passwords are lost. It presents a practical, model vouching system reliant on the popular equipment confirmation token SecurID. In addition to standard cryptographic security requirements, it also considers social design and customer behavior questions [14].

Damopoulos et al. [15] suggested in their paper that the proliferation of touchscreen devices presents some intriguing exploratory problems. Among these is the question of whether touch stroke-based analysis (like keylogging) can serve as a reliable means of characterizing a connected device's user base. It goes without saying that the coin has two sides in such a situation. First, the yield produced by this kind of framework can be used to manage AI classifiers, and subsequently, interruptive discovery motors. Second, to obtain private client data, aggressors may install touch loggers. In the past, heritage keyloggers have also widely used this malicious option in many contexts; nevertheless, it has hardly been tested for sensitive consoles. Between these opposing but complementary perspectives, this research implements the first-ever local and fully functioning touch logger for modern smartphones, especially those using the closed-source iOS platform. This technique resulted in a very positive appearance and accuracy in identifying abuses, and as a result, post-validating the client, in an amount that exceeds about 100%, for the primary purpose. Real-world use cases also demonstrate how harmful such programming can be when used in a vengeful manner [15].

In ref. [16], a blockchain-based new mechanism is proposed, which provides hybrid computing, decentralized electronic health records, and smart-contract-based service automation. The proposed system uses hybrid computing and blockchain-based data storage, ensuring security, privacy, and low latency. The analysis shows that this system is capable of meeting the requirements of the decentralized IoMT smart healthcare system.

Soofi et al. [17] discussed that the mechanism for safeguarding information is crucial, otherwise there are chances of theft, hence privacy is at risk. In IoT-enabled devices security and privacy are important.

## 6.3 Proposed System

### 6.3.1 System architecture and workflow

In this work, we propose an advanced authentication framework that enhances both security and user experience, specifically for online transactions. The system eliminates static usernames and passwords by implementing a one-time username (OTU) and one-time password (OTP), along with additional layers of cryptographic security.

The proposed framework can be utilized in various areas, for example, the power bill and the water bill. The web-based financial framework is used to exhibit the convention. We can make a one-time secret phrase with the check of the picture to make the instalment. To make the bill installment, the help and the card type, then the card holder name alongside the card number and expiry date, with the CVV number is utilized. To make the instalment OTP, alongside the chosen picture is utilized to give the safe verification. In the server, the client subtleties the key age subtleties alongside the bill instalment, and the unauthenticated subtleties can be seen (Figure 6.3). The proposed framework defines different steps to ensure the user's security and smooth payment processing.

User initiation: The process begins when the user logs in, during which a unique and one-time username (OTU) and password are generated, which is valid for that session only. It is a crucial step as it does not require a static username and password.

Payment details input: Once logged in, the user enters payment details, such as card details, CVV number. This information is transmitted to the server through a secure channel.

Image verification and OTP: The framework further enhances security through image-based authentication and one-time password (OTP) verification. The user is required to select an image they have pre-determined for better security concerns, and a OTP is sent to the user for verification. The combination of image selection and OTP provides robust security authentication.

User details storage, key generation, and bill payment processing: Once the server receives user information and payment details, it stores the user details.

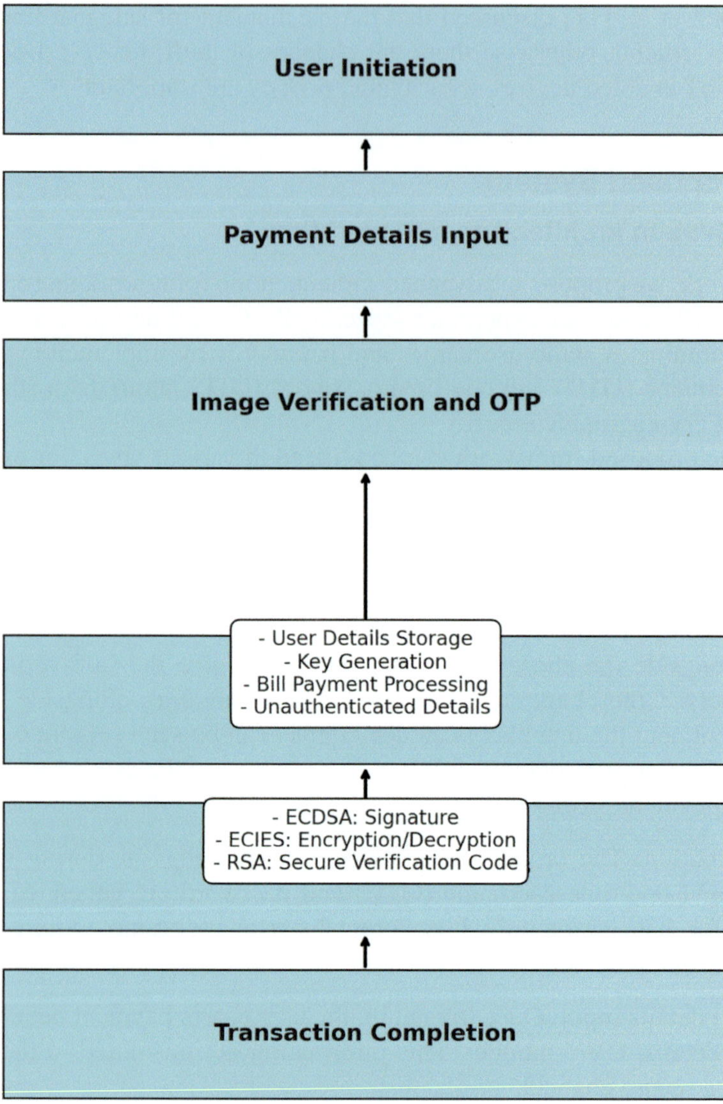


**Figure 6.3** Proposed system.

This is followed by the key generation process using digital signatures and encryption techniques. Afterward, the bill payment process is completed, and the payment is successfully processed. Only authenticated information is stored on the server, thereby preventing unverified details from being used at any time.

The system employs various cryptographic techniques for security. ECDSA (elliptic curve digital signature algorithm) is used for digital signatures, ECIES (elliptic curve integrated encryption scheme) is utilized for encryption and decryption, and the RSA algorithm ensures data privacy and verification. The Asymmetric cryptosystem that uses the RSA calculation to ensure the privacy of the confirmation code.

In the end, after successful completion of all steps, the user receives confirmation of the payment. The system ensures that all information is secure and inaccessible to unauthorized individuals.

### 6.3.2 Security analysis and attack mitigation

In this part, we illustrate how the security protocols and cryptographic techniques we use in our work might thwart various attacks that aim to compromise the suggested validation process. Additionally, we will analyze how the system effectively counters common threats like phishing attacks, password-based vulnerabilities, replay attacks, and shoulder-surfing attacks. We expect the enrolled device to be in a secure environment where cryptographic computations can be performed [18]. The payment card industry data security standard, or PCI DSS, is something we also need the verification server to obtain and fully comply with.

#### 6.3.2.1 Protection against phishing attacks

Using the appearance of a trustworthy source, phishing attacks aim to get credentials such as usernames and secret phrases. The validation that has been suggested can help reduce the risk associated with phishing attacks. Since the suggested method lacks a set username or secret word, it is, in fact, hostile to phishing techniques. There is always a fresh username and secret phrase for each confirmation meeting. For the reasons listed below, we view the suggested plan as hostile to the phishing validation standard. As part of the current work, a login is created for the client to use during a single meeting; this username is significant only temporarily. For confirmation, we use a secure channel, and for approval, we use a client-driven admission control. There is no comprehension of username or secret key update in the present work [19].

#### 6.3.2.2 Resistance to attacks related to passwords

The suggested setup provides protection against many attacks linked to secret words, such as direct perception attacks and shoulder-riding attacks. At present, the client is not allowed to utilize static usernames and passwords, which can be identified by mechanical vibration analysis or thermal imaging to determine which keys were pushed. The secret phrase and username change with each login session, so we can prevent issues like utilizing the customer's birthday as the secret phrase, using the same secret phrase everywhere, or forgetting the secret key [20].

#### 6.3.2.3 Defense against shoulder surfing attacks

Another type of attack that occurs when someone uses a static username and secret word combination is shoulder-riding, which is typically employed to obtain sensitive information, such as the secret word. A toxic abuser using various direct perception techniques identifies the target and obtains certifications. Examining the victim's shoulder for a hidden key is one straightforward tactic. With the use of vision-improving devices, shoulder-riding attacks may even be carried out from a considerable distance. To prevent shoulder surfing, we use dynamic qualifications that are established and used just once, depriving aggressors of any value when collecting login data from victims [21].

#### 6.3.2.4 Replay attack mitigation

To evade verification requests, customers must refresh their meeting key and one-time login. The once-use secret phrase also expires after usage or after an incredibly short period of time. Accompanying the user login list ULL with time stepping provides a potent way to prevent a repeat attack. Note that the customer's personality is verified only once, for a very short period (like 5 min), by the server, who generates a confirmation code. Even if an attacker intercepts a request, the one-time nature of our authentication process prevents reuse. We can assure the server's ability to fend off replay attacks in this manner.

#### 6.3.2.5 Secure client-server communication

Under our strategy, a mark is applied to a customer's validation request, which ensures that the messages authenticity and information respectability. The public key of the server, which is reliant on ECIES, is used to encrypt the customer's confirmation request and mark. In the irregular prophet model, ECIES is provably secure, ensuring the secrecy of the marked and encoded communications.

To get the customer's request and record it as soon as it receives it with the one-time secret key data, the server decodes the message. The server confirms the client's identification at that point. The server waits for the client to log in and discards the one-time secret phrase if the client is not authorized. After logging in with OTU, the client gets the scrambled check code from the server. The confirmation code is generated by the server and encrypted with the meeting key [22]. Since *k* is the common meeting key that only the customer and the server know, only the customer may decode the encoded check code to retrieve the confirmation code once they have received it. Therefore, the privacy of the response message is ensured throughout the reaction procedure. By implementing the security measures: ECIES-256 encryption for protecting data confidentiality, RSA-based digital signatures, and AES-128 encryption for authentication tokens, we guarantee data integrity and protect against man-in-the-middle (MITM) attacks.

### 6.3.3 System modules and threat mitigation

Registration Module: The client must register a smart device with the server in this module. An excellent personal device, like a cell phone, is an enrolled device. It ought to be able to carry out cryptographic operations. For each client to use the server's services, they must enroll a device with the server. There should be no need for a static username and secret key for a valid client to access server functions.

Server Module: This module requires the server to verify the customer's information. The server is connected to an equipment security module and has a location with a substance, like a bank. RSA, which provides crypto-handling and protects the private key. Along with providing various forms of support, the server gives clients access to its public key and confirmation code.

Consumer Module: The client terminal must input the client's name that was sent by the enrolled device once in this module. A client's terminal, which is an electronic device such as a PC or workstation, is used to log in to the server to see or conduct trades. The consumer module integrates dynamic credentials, preventing replay and shoulder-surfing attacks.

## 6.4 Experimental Setup and Results

In terms of border and ciphertext sizes, we break down the correspondence overhead. We choose a size of 4 bytes for the entrance control field, 128 bytes for the meeting key k, and 8 bytes for the one-time username (OTU). Furthermore, the timestamp is 32 bytes, and the TVP (one-time secret key validity period) size is set to 4 bytes (Figure 6.4).

For privacy protection, we choose the ECIES-256 cryptosystem; for message signing, we use a further ECDSA-256 cryptosystem; and for check code security, we use the AES 128-cycle cryptosystem.

| **Client Operations** | **Server Operations** |
|---|---|
| *Step1. Generates a random session key:*<br>*k="y9Q7A....32C3"* | *Step1: Decrypts Encr_k to get k:*<br>*k'= Decrypt (Enr_k, PrivateKey_Server)* |
| *Step2. Signs k using ECDSA-256:*<br>*σ = Sign (k, PrivateKey_Client)* | *Step2: Verifies sigma using ECDSA-256:*<br>*Verify(σ, k', PublicKey_Client)* |
| *Step3: Encrypts k using ECIES-256:*<br>*Encr_k=Encrypt(k, PublicKey_Server)* | *Step3: If verification succeeds, the session is authenticated.* |
| *Step4: Sends {Encr_k, σ} to Server* | |

**Figure 6.4** Illustration of secure client-server communication.

We now compute the total computational cost of the following operations separately:

- ECDSA-256 signature generation → $C_{sn}$
- ECDSA-256 signature verification → $C_{ve}$
- ECIES-256 encryption → $C_{en}$
- ECIES-256 decryption → $C_{de}$

The computational overhead from the client side is the sum of the signing cost and the encryption cost, that is, $C_{sn} + C_{en}$. The computational overhead from the server side is the sum of the decryption cost and the verification cost, that is, $C_{de} + C_{ve}$ (Table 6.1).

The computational costs associated with ECDSA-256 and ECIES-256 are well documented. ECDSA-256 offers comparable security to much larger RSA keys, but with reduced computational overhead, making it efficient for both client and server operations. Our approach ensures minimal processing delay while maintaining robust security.

**Table 6.1** Parameters for security.

| **Operation** | **Client side** | **Server side** |
|---|---|---|
| Encryption | $C_{en}$ | 0 |
| Decryption | 0 | $C_{de}$ |
| Signature | $C_{sn}$ | 0 |
| Verification | 0 | $C_{ve}$ |
| Computational over head | $C_{en} + C_{sn}$ | $C_{de} + C_{ve}$ |

Compared to current authentication methods, the proposed system has a number of advantages. It uses dynamic one-time usernames and OTPs, which improve security, in contrast to conventional systems that use static usernames and passwords. The suggested solution incorporates cutting-edge cryptographic algorithms, including ECIES-256 encryption and ECDSA-256 digital signatures for enhanced data protection, whereas current systems rely on password hashing. By using time-stamped one-time credentials rather than just session timeouts, it also lessens replay attacks. The system uses OTPs and image-based verification in place of static password input to prevent shoulder surfing. In comparison to RSA-based encryption, the suggested method lowers costs by optimizing computing performance through the use of elliptic curve cryptography-based cryptographic operations. Finally, by using dynamic credentials for seamless authentication and doing away with the need to learn passwords, it improves user convenience.

## 6.5 Conclusion

Due to the phenomenal rise of online banking and online business frameworks, individual clients now manage a huge number of usernames and passwords. Various security vulnerabilities affect standard static username and secret key protocols. Many customers start repeatedly using duplicate certificates in various documents and systems. A perpetrator may breach multiple frameworks and endanger the safety and security of their clients by disclosing or compromising one record. Here, we offer an additional validation model that enables users to resolve a variety of problems, such as forgetting login credentials for numerous websites and frameworks. A client-driven admission control system that lowers the risk of numerous assaults is facilitated by the proposed verification plot.

By leveraging advanced cryptographic techniques, the proposed model ensures robust privacy protection. The computational load in this research is split between the client and the server, where the client focuses on signature generation ($C_{sn}$) and encryption ($C_{en}$) , whereas the server looks after decryption ($C_{de}$) and verification ($C_{ve}$). Though the approach is optimized, the use of advanced cryptographic techniques like ECIES-256 and ECDSA-256 may still impose high computational costs on resource-constrained devices, such as mobile phones or IoT devices. Latency and bottlenecks during high traffic can be observed as a challenge. In future work, fraud detection models using machine learning can be introduced to analyze login patterns.

## References

[1] H. Hui, C. Zhou, S. Xu, and F. Lin, "A novel secure data transmission scheme in industrial Internet of Things," *China Communications*, vol. 17, no. 1, pp. 73-88, 2020.

[2] K. Tange, M. De Donno, X. Fafoutis, and N. Dragoni, "A systematic survey of industrial Internet of Things security: Requirements and fog computing opportunities," *IEEE Communications Surveys & Tutorials*, vol. 22, no. 4, pp. 2489-2520, 2020.

[3] M. Serror, S. Hack, M. Henze, M. Schuba, and K. Wehrle, "Challenges and opportunities in securing the industrial Internet of Things," *IEEE Transactions on Industrial Informatics*, vol. 17, no. 5, pp. 2985-2996, 2021.

[4] A. A. Khan et al., "BIoMT: A state-of-the-art consortium serverless network architecture for healthcare system using blockchain smart contracts," *IEEE Access*, vol. 10, pp. 78887-78898, 2022.

[5] Ernawati, T., Suhardi, & Nugroho, D.R. "IT risk management framework based on ISO 31000:2009" 2012 International Conference on System Engineering and Technology (ICSET), pp. 1-8, 2012

[6] Y. Liu et al., "Blockchain bridges critical national infrastructures: E-healthcare data migration perspective," *IEEE Access*, vol. 10, pp. 28509-28519, 2022.

[7] U. Hariharan, V. Dhanakoti, K. Rajkumar, and J. Jeyavel, "Safeguarding e-healthcare documents utilizing an infrastructure for advanced keyless signature infrastructure blockchain within the cloud," in *Proc. Int. Conf. Emerging Smart Computing and Informatics (ESCI)*, 2022, pp. 1-6.

[8] P. Kumar et al., "A blockchain-orchestrated deep learning approach for secure data transmission in IoT-enabled healthcare system," *J. Parallel Distrib. Comput.*, vol. 172, pp. 69-83, 2023.

[9] Y. Alemami, A. M. Al-Ghonmein, K. G. Al-Moghrabi, and M. A. Mohamed, "Cloud data security and various cryptographic algorithms," *Int. J. Electr. Comput. Eng. (IJECE)*, vol. 13, no. 2, pp. 1867-1879, 2023.

[10] A. Alrawais, C. Alhothaily, X. Hu, X. Xing, and X. Cheng, "An attribute-based encryption plan to get mist interchanges," *IEEE Access*, vol. 5, pp. 9131-9138, 2017.

[11] J. Bonneau, C. Herley, P. C. Van Oorschot, and F. Stajano, "The journey to supplant passwords: A system for near assessment of web validation plans," in *Proc. IEEE Symp. Security & Privacy (SP)*, 2012, pp. 553-567.

[12] K. L. Neela and V. Kavitha, "Enhancement of data confidentiality and secure data transaction in cloud storage environment," *Cluster Computing*, vol. 21, no. 1, pp. 115-124, 2017.
[13] M. Borchert and M. Gunther, "Indirect NFC-login," in *Proc. 8th Int. Conf. Web Technol.*, 2013, pp. 204-209.
[14] J. Brainard, A. Juels, R. L. Rivest, M. Szydlo, and M. Yung, "Fourth-factor verification: Somebody you know," in *Proc. 13th ACM Conf. Computer and Communications Security*, 2006, pp. 168-178.
[15] G. Damopoulos, G. Kambourakis, and S. Gritzalis, "From keyloggers to touchloggers: Take the harsh with the smooth," *Computers & Security*, vol. 32, pp. 102-114, 2013.
[16] B. S. Egala, A. K. Pradhan, V. Badarla, and S. P. Mohanty, "Fortified-chain: A blockchain-based framework for security and privacy-assured internet of medical things with effective access control," *IEEE Internet Things J.*, vol. 8, no. 14, pp. 11717-11731, 2021.
[17] A. A. Soofi and H. Ahmad, "Deep learning for IoT security: Applications and challenges," in *Emerging Technologies for Securing the Cloud and IoT*, 2024, pp. 69-99.
[18] A. Das, J. Bonneau, M. Caesar, N. Borisov, and X. Wang, "The tangled snare of secret word reuse," in *Proc. Symp. Netw. Distrib. Syst. Security (NDSS)*, 2014, p. 1.
[19] X. Tooth and J. Zhan, "Online financial verification utilizing incorporated telephones," in *Proc. 5th Int. Conf. Future Inf. Technol. (FutureTech)*, 2010, pp. 1-5.
[20] Y. S. Lee, N. H. Kim, H. Lim, H. Jo, and H. J. Lee, "Online financial verification framework utilizing incorporated OTP with QR-code," in *Proc. 5th Int. Conf. Comput. Sci. Convergence Inf. Technol. (ICCIT)*, 2010, pp. 644-648.
[21] M. Mannan and P. C. Van Oorschot, "Passwords for both incorporated and PCs: OBPWD for Refox and Android," *USENIX; Login*, vol. 37, no. 4, pp. 28-37, 2012.
[22] M. Mannan and P. C. Van Oorschot, "Leveraging individual gadgets for more grounded secret word validation from untrusted PCs," *J. Comput. Security*, vol. 19, no. 4, pp. 703-750, 2011.
[23] M. Pathak, K. N. Mishra, and S. P. Singh, "Securing data and preserving privacy in cloud IoT-based technologies: An analysis of assessing threats and developing effective safeguards," *Artif. Intell. Rev.*, vol. 57, no. 10, p. 269, 2024.
[24] V. Sharma, A. Kero, H. C. Sharma, and P. Semwal, "A study on security of IoT: Problems solving using ML and blockchain," *J. Pharm. Negative Results*, pp. 1311-1323, 2023.

[25] S. Upadhyay, R. Roy, and S. Ghosh, "Designing hash and encryption engines using quantum computing," in *Proc. 37th Int. Conf. VLSI Design and 23rd Int. Conf. Embedded Syst. (VLSID)*, 2024, pp. 571-576.

[26] S. Bharati and P. Podder, "Machine and deep learning for IoT security and privacy: Applications, challenges, and future directions," *Security Commun. Netw.*, vol. 2022, p. 8951961, 2022.

[27] I. H. Sarker, A. I. Khan, Y. B. Abushark, and F. Alsolami, "Internet of things (IoT) security intelligence: A comprehensive overview, machine learning solutions, and research directions," *Mobile Networks and Applications*, vol. 28, no. 1, pp. 296-312, 2023.

[28] N. Mishra and S. Pandya, "Internet of things applications, security challenges, attacks, intrusion detection, and future visions: A systematic review," *IEEE Access*, vol. 9, pp. 59353-59377, 2021.

[29] L. Cambosuela, M. Kaur, and R. Astya, "The vulnerabilities and risks of implementing Internet of Things (IoT) in cybersecurity," in *Proc. 2024 11th Int. Conf. Reliability, Infocom Technol., and Optimization (Trends and Future Directions) (ICRITO)*, 2024, pp. 1-5.

[30] M. A. Amanullah et al., "Deep learning and big data technologies for IoT security," *Comput. Commun.*, vol. 151, pp. 495-517, 2020.

[31] A. A. Khan, A. A. Shaikh, Z. A. Shaikh, A. A. Laghari, and S. Karim, "IPM-model: AI and metaheuristic-enabled face recognition using image partial matching for multimedia forensics investigation with genetic algorithm," *Multimedia Tools and Applications*, vol. 81, no. 17, pp. 23533-23549, 2022.

[32] W. Shen, J. Qin, J. Yu, R. Hao, and J. Hu, "Enabling identity-based integrity auditing and data sharing with sensitive information hiding for secure cloud storage," *IEEE Trans. Inf. Forensics Secur.*, vol. 14, no. 2, pp. 331-346, 2019, doi: 10.1109/TIFS.2018.2850312.

[33] R. Li et al., "A lightweight secure data sharing scheme for mobile cloud computing," *IEEE Trans. Cloud Comput.*, vol. 6, no. 2, pp. 344-357, 2018, doi: 10.1109/TCC.2017.2649685.

[34] J. Li et al., "An efficient attribute-based encryption scheme with policy update and file update in cloud computing," *IEEE Trans. Ind. Informat.*, vol. 15, no. 12, pp. 6500-6509, 2019, doi: 10.1109/TII.2019.2931156.

[35] Y. Li et al., "Fuzzy identity-based data integrity auditing for reliable cloud storage systems," *IEEE Trans. Dependable Secure Comput.*, vol. 16, no. 1, pp. 72-83, 2019, doi: 10.1109/TDSC.2017.2662216.

[36] L. Zhang, Y. Cui, and Y. Mu, "Improving security and privacy attribute-based data sharing in cloud computing," *IEEE Syst. J.*, vol. 14, no. 1, pp. 387-397, 2020, doi: 10.1109/JSYST.2019.2911391.

[37] S. Wang et al., "An efficient file hierarchy attribute-based encryption scheme in cloud computing," *IEEE Trans. Inf. Forensics Secur.*, vol. 11, no. 6, pp. 1265-1277, 2016, doi: 10.1109/TIFS.2016.2523941.

[38] W. Teng, G. Yang, Y. Xiang, T. Zhang, and D. Wang, "Attribute-based access control with constant-size ciphertext in cloud computing," *IEEE Trans. Cloud Comput.*, vol. 5, no. 4, pp. 617-627, 2017, doi: 10.1109/TCC.2015.2440247.

[39] W. Li, K. Xue, Y. Xue, and J. Hong, "TMACS: A robust and verifiable threshold multi-authority access control system in public cloud storage," *IEEE Trans. Parallel Distrib. Syst.*, vol. 27, no. 5, pp. 1484-1496, 2016, doi: 10.1109/TPDS.2015.2448095.

[40] X. Ma, J. Ma, H. Li, Q. Jiang, and S. Gao, "PDLM: Privacy-preserving deep learning model on cloud with multiple keys," *IEEE Trans. Services Comput.*, vol. 14, no. 4, pp. 1251-1263, 2021, doi: 10.1109/TSC.2018.2868750.

[41] I. Gupta and A. K. Singh, "A confidentiality preserving data leaker detection model for secure sharing of cloud data using integrated techniques," in *Proc. 7th Int. Conf. Smart Comput. & Commun. (ICSCC)*, 2019, pp. 1-5, doi: 10.1109/ICSCC.2019.8843648.

# 7

# Optimizing Mental Health Interventions Through AI-Based Therapies

**Harendra Singh Negi[1], Bhawnesh Kumar[2] and Aditya Joshi[3]**

Department of Computer Science and Engineering, Graphic Era Deemed to be University, India

**Email:** mail.harendrasinghnegi@gmail.com; bhawneshmca@gmail.com; adi.joshi@geu.ac.in

## Abstract

Artificial intelligence-enabled psychological treatment refers to innovative approaches that utilize artificial intelligence to provide psychological assistance. This model leverages natural language processing, machine learning, and sentiment analysis to simulate human-like therapy conversations, delivering personalized and immediate psychological support. By integrating psychological counselling, guided meditations, emotional support, and behavioral monitoring, AI-driven psychological treatment offers a range of interventions that are available at any time. Through text or voice interactions, the system analyzes user inputs to assess emotions, detect psychological distress, and identify underlying concerns. Considering the individual's emotional state, mental well-being, and past experiences, the artificial intelligence generates tailored therapeutic responses, including coping strategies, exercises, and counselling techniques. As users engage with the system over time, it continuously adapts and refines its interventions, improving effectiveness and ensuring more personalized support. By offering a confidential, adaptable, and cost-effective solution for psychological support, this artificial intelligence-driven approach helps overcome the challenges associated with traditional mental healthcare. Additionally, it provides clients with flexible options to attend therapy sessions, potentially increasing treatment adherence



DOI: 10.1201/9788743812739-7

and demand. The model continuously enhances its responses by integrating feedback surveys and behavioral data analysis, ensuring therapy remains relevant and effective. Secure connections and adherence to medical regulations safeguard user confidentiality and privacy. Ultimately, artificial intelligence-powered psychological treatment has the potential to revolutionize mental healthcare by making therapy more scalable, personalized, and accessible, while providing continuous support to individuals facing psychological challenges.

## 7.1 Introduction

One of the most essential expansions in the recent change of the healthcare sector is psychological treatment. Historically reserved for patients and physicians, technology is now a third-party present in the sector. ICT is currently a primary aspect and the industry experts in patient-level healthcare, despite having already established itself in the medical field and its operations [1, 2]. These days, wearable technology and virtual reality (VR) are used to meet the goals of individualized interactive training with increased patient motivation [3, 4]. Despite their effects, both technologies are enabling themselves to have a bigger presence and influence on the lives of doctors and patients through gadgets that are always with them. As a result of technological advancements, mobile apps for healthcare are becoming a crucial component of users' home screens [5, 6]. When combined with revolutionary technologies like virtual reality and wearable medical equipment, these apps have the potential to drastically alter how the health ecosystem engages with the general public and are growing in the healthcare industry. Together, the mobile devices and mobile applications serve as a platform for communication, computing, and a Graphical User Interface (GUI) for both clinical staff and patients [7, 8]. Improving the quality of physical therapy services and lowering costs in light of the strain that aging phenomena are placing on the system are two of the driving forces behind this research [9, 10]. The normal process of aging is typically accompanied by a decline in everyday motor function linked to muscle-skeletal deficits, which can also happen following a stroke. Another demographic that needs physical rehabilitation therapy to enhance their motor abilities is post-stroke patients [11, 12]. According to the data, stroke is the leading cause of serious physical disability worldwide. According to estimates, there will be between 7 and 8 million stroke fatalities and 23 million first-ever strokes in 2030 if secular trends continue [13]. Strokes typically affect one side of the body and can result in significant physical problems such as paralysis, discomfort, weakness, and attention deficit. The inability to carry out daily tasks may be the consequence of such deficits [14]. There are a number of approaches that can be developed to help stroke sufferers live better lives [15]. During periods of constant economic and technological advancement, many people experience sadness brought on, among other things, by the necessity of attending rehabilitation sessions, as well as physical and/or motor restrictions for a variety of causes. Scientists are always searching for innovative methods to prevent and treat physical diseases, as well as ways to use the most recent advancements in computer technology to intervene in diagnosis to make it more successful and effective for individuals in need of long-term care. Therapeutic techniques based on various virtual reality scenario applications have garnered significant interest from the health community [16]. Serious games are becoming more and more integrated into various organizations, including the military, healthcare, education, and entertainment, thanks to advancements in game production and computer graphics hardware. In physical therapy, virtual reality (VR) has been utilized to help stroke victims practice upper limb mobility [17, 18, 19]. Numerous virtual reality games pertaining to physical rehabilitation have been documented in the literature [20]. During in-person training sessions, the games enable patients to engage with virtual objects in real-time, enhancing their specific motor abilities. Setting up the system for patients in accordance with specific prescriptions is the responsibility of the therapists. However, the market's supply of body measuring sensors in wearable form has grown exponentially, enabling customers to develop monitoring applications ranging from basic heart rate monitoring to limb movement response. These sensors produce various datasets that are instantly accessible for the researchers' various data-finding methods, such as in the feedback process. Applications in healthcare are found to benefit greatly from the significant advancements in networking technology and measurement sensor capabilities [21]. IoT healthcare solutions are already evident on a worldwide level, as new IoT-compatible devices are replacing outdated technology and offering the opportunity to create and implement innovative, useful techniques as part of healthcare services. Wearable technology has the potential to completely transform healthcare services by monitoring physiological and motor parameters and providing individualized treatment based on data about behavior, physical activity, health status, and other factors like air quality that impact daily living. To create a scalable virtual therapy system that bridges the gap in mental healthcare by offering accessible, efficient, and private support for individuals seeking emotional and psychological well-being.

## 7.2 Related Work

Wearable technology has emerged as a highly practical way to monitor a patient's health. Medical wearables that not only help monitor but also treat and rehabilitate chronic diseases are becoming more and more common. Examples include devices that use electro-muscular stimulation to encourage muscle activation and heal injuries [22], devices that use ECG and EMG biosensors to detect cardiovascular diseases [23], and pressure and force sensors that are placed in insoles and linked to software tools (web or mobile apps) to analyze and track a patient's gait. Numerous research studies are being conducted that offer novel approaches to integrating VR gaming tools and new technologies from the IoT industry to develop new healthcare solutions that emphasize physical rehabilitation [24, 25]. Games' ability to immerse players and measure different kinds of data using wearable computers and tiny sensors has been seen as a great way to address the difficulties associated with physical rehabilitation. Virtual reality (VR), augmented reality (AR), mixed reality (MR), sensors, telecommunications technologies, computer interfaces, dedicated servers, and cloud services are only a few of the technologies that are used in serious gaming. During motor rehabilitation, these technologies can help capture complicated kinetic and kinematic variables (such as walking or standing pressure, limb movement duration, and speed) with accuracy and detail [26]. With the help of visual and aural incentive elements, virtual reality rehabilitation systems can transform the conventionally dull rehabilitation exercises into enjoyable ones. Furthermore, these elements are pre-programmed to support patients by generating auditory or visual motivating messages in reaction to potential triggers from the patient. Events are categorized according to the patient's potential progress in the corresponding exercise. Events are defined by a therapist according to the exercise situation. With the help of visual and aural incentive elements, virtual reality rehabilitation systems can transform the conventionally dull rehabilitation exercises into enjoyable ones [27]. Furthermore, these elements are pre-programmed to support patients by generating auditory or visual motivating messages in reaction to potential triggers from the patient. Events are categorized according to the patient's potential progress in the corresponding exercise. Events are defined by a therapist according to the exercise scenario [28]. A previous study analyzed the conditions of brain neurons and observed their gradual recovery after 6 weeks, ultimately enabling the subjects to slowly control leg movements, indicating that long-term VR stimulation benefits limb rehabilitation and training [29]. Another study combined VR with neurophysiology to stimulate the participants' brains through localized induction to facilitate hand movements [30]. The research has shown that VR-assisted rehabilitation improves neural conduction speed in the arms faster than traditional methods, leading to enhanced upper limb recovery [31]. Thayer et al. suggest that autonomic dysfunction and impaired neurovascular coupling in ischemic stroke may contribute to further brain injury and cognitive deterioration [32]. The key innovation of this study lies in the seamless integration of the EEG technology and real-time compensatory detection. In the future, they need to conduct more extensive clinical trials to validate the efficacy and feasibility of the system across different clinical scenarios. As mentioned in previous studies, this valence-arousal emotional two-dimensional space includes regions of fear, joy, disappointment, pleasure, sadness, and satisfaction that are usually used in psychology [33, 34]. We integrate this into our VR rehabilitation system. A patient's emotions can be identified based on the position of the mapped point, aiding the therapist in understanding the patient's current psychological state. Simultaneously, through a combination of EEG technology and VR, our VR rehabilitation system can quantify a patient's current level of positivity and can assist therapists in tailoring treatment according to the patient's positivity level [35].

## 7.3 Problem Statement and Methodology

Mental health conditions like anxiety, sadness, stress, and other emotional difficulties are becoming more common in today's fast-paced society. However, access to mental healthcare remains limited due to barriers such as geographical location, long wait times, stigma, and the high cost of traditional therapy services. Many individuals either do not seek professional help or cannot receive it regularly, which exacerbates mental health problems. The problem is the need for an accessible, affordable, and confidential solution that allows individuals to receive emotional support, therapeutic guidance, and mental healthcare. There is a growing demand for scalable and reliable mental health interventions that can complement traditional therapy or serve as an initial point of contact for those in need, and the challenges that need to be addressed are shown in Figure 7.1.

A system for psychologists that gives quick, individualized psychological treatment through the use of artificial intelligence and natural language processing. While ensuring privacy and security, the platform will provide clients with behavioral monitoring, meditation exercises, survival strategies, and methods to alleviate stress . Establishing an efficient platform for psychological treatment needs AI techniques that may somewhat replicate

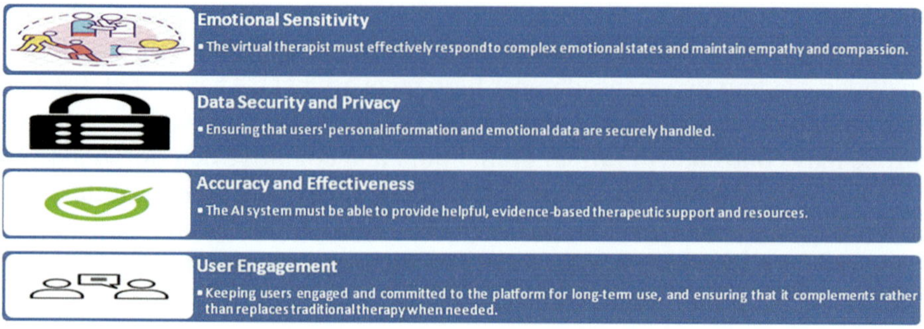


**Figure 7.1** Challenges in virtual therapy.

the therapeutic and sympathetic connections seen in conventional human treatment. The AI system needs to be able to comprehend normal language, respond in a personalized way, and gradually modify its strategy based on the user's emotional state. The main AI techniques that can be used to create a virtual therapist are broken down below:

### 7.3.1 Natural language processing (NLP)

NLP forms the backbone of communication in virtual therapy, allowing the system to understand, interpret, and respond to human language effectively. The system needs to analyze both text and speech (if voice-based therapy is included).

- **Sentiment Analysis:** NLP algorithms can detect the emotional tone of a user's input (e.g., happiness, sadness, anxiety, frustration), allowing the virtual therapist to respond empathetically and appropriately.
- **Intent Recognition:** The system identifies the user's underlying intent (e.g., whether they're seeking advice, sharing a problem, or simply needing to vent) to provide contextually relevant responses.
- **Emotion Detection:** Using techniques like text classification and deep learning, the virtual therapist can understand specific emotional states (e.g., depressive, anxious, angry), helping it choose the best therapeutic approach (e.g., cognitive-behavioral therapy, mindfulness, or motivational interviewing).
- **Contextual Understanding:** The system should maintain context over multiple interactions, remembering past conversations and long-term mood trends, which is essential for providing personalized therapy.

### 7.3.2 Reinforcement learning (RL)

Reinforcement learning can be used to optimize the virtual therapist's interactions with users. The system learns from user feedback and interactions to improve its performance over time.

- **Reward Mechanism:** Positive responses or progress (e.g., user reports feeling better after an exercise) act as a "reward," reinforcing the virtual therapist's approach.
- **Exploration vs. Exploitation:** The system must balance trying new therapeutic techniques (exploration) with sticking to proven methods that have worked for the user (exploitation).
- **Personalized Therapy:** RL can be used to fine-tune responses and treatment plans for each individual, based on ongoing interactions and feedback. For example, the system may learn that a user responds better to certain techniques (e.g., mindfulness exercises) than others (e.g., problem-solving methods).

### 7.3.3 Machine learning (ML) for personalization

The virtual therapist can leverage machine learning algorithms to tailor its responses to the individual user's needs, preferences, and mental health history.

- **User Profiling:** The system builds a profile of each user, including emotional history (e.g., past stressors, recurring issues), preferences (e.g., preferred therapy methods), and progress over time. This profile can be updated based on each interaction, allowing the system to adapt its approach accordingly.
- **Recommendation Systems:** Just like in e-commerce or entertainment platforms, a recommendation engine can suggest specific resources (e.g., articles, exercises, meditation guides) or therapeutic techniques (e.g., CBT, DBT) based on the user's needs and progress.
- **Behavioral Prediction:** Machine learning can be used to predict potential changes in the user's mental state, such as identifying patterns that may signal an emerging mental health crisis (e.g., escalating anxiety or depressive tendencies). Early detection allows the system to intervene and offer additional support.

### 7.3.4 Speech recognition and conversational AI

If the virtual therapist includes a voice-based interface, integrating advanced speech recognition and conversational AI models will be key to facilitating real-time, human-like interactions.

- **Speech-to-Text (STT):** Converting spoken input into text that the system can understand, using models like Google speech-to-text or deep learning-based speech recognition.
- **Conversational Agents:** AI models like GPT-4, combined with structured therapy frameworks (e.g., CBT, narrative therapy), will guide the conversation. The system should simulate empathy, adjust its tone based on the emotional state, and maintain a helpful and non-judgmental tone.

By combining these AI methodologies, the virtual therapist can offer users a personalized, effective, and compassionate mental health support system. The key challenge is ensuring that the system is not only intelligent and efficient but also ethical, empathetic, and capable of providing value to individuals with a wide range of mental health needs.

## 7.4 System Implementation

### 7.4.1 System architecture

Below is a block diagram representing the system architecture for a virtual therapy platform. The diagram illustrates the key components and their interactions in a visual manner, showing the flow of data and processes within the system.

The block diagram shown in Figure 7.2 provides a high-level overview of how the components of the virtual therapy system interact and function together to provide real-time, personalized, and secure therapeutic support to users. Each layer and block represents a crucial part of the system, from user interaction to AI-driven therapy, data storage, and compliance.

### 7.4.2 Workflow diagram

A workflow diagram visually represents the sequence of processes and interactions between the user and the system in a Virtual Therapy platform. The workflow describes the steps from the user's initiation of a therapy session to the final outcomes, including AI processing, session management, and feedback collection.

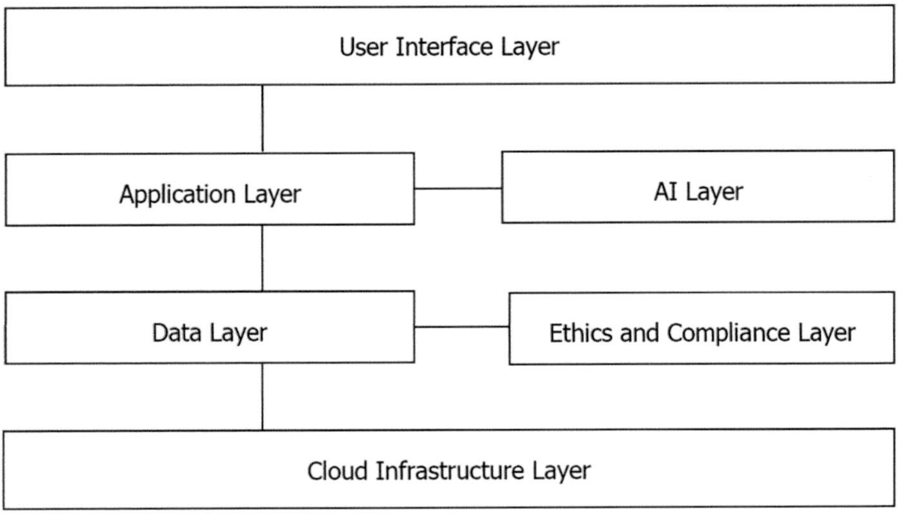


**Figure 7.2** System architecture.

The workflow diagram shown in Figure 7.3 outlines a step-by-step sequence of events and interactions that occur during a virtual therapy session. It provides a comprehensive overview of the user experience, from logging in to engaging with the virtual therapist, receiving personalized therapy, and tracking progress over time. This diagram helps in understanding the end-to-end process flow of a virtual therapy platform.

### 7.4.3 Proposed work

A proposed work diagram for virtual therapy outlines how the system processes and integrates different components to deliver a therapeutic experience

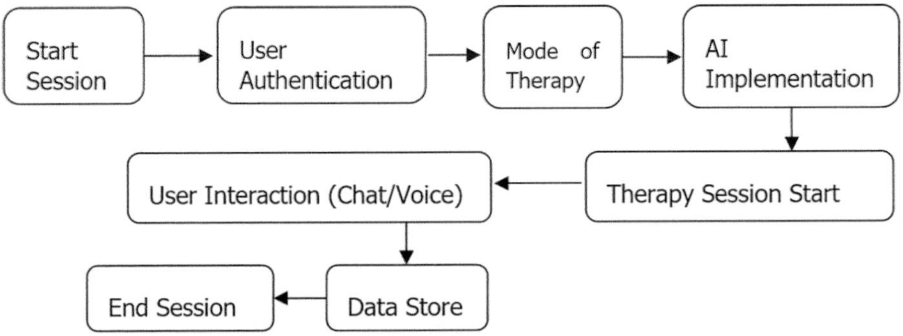


**Figure 7.3** Workflow diagram.

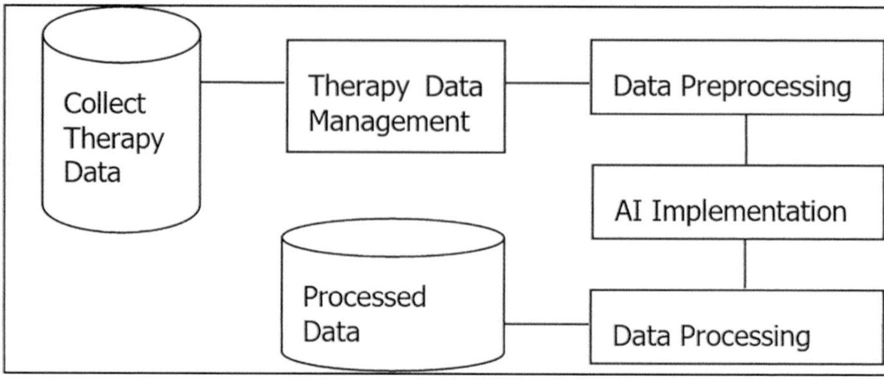


**Figure 7.4** Proposed work for virtual therapy.

for users. This includes data collection, data flow, AI processing, and system responses. The diagram represents the key stages involved in providing therapy through AI (virtual therapist) and emphasizes the data flow between various modules.

This proposed work diagram, shown in Figure 7.4 illustrates the end-to-end process of a virtual therapy session, detailing how data collection, AI processing, and therapeutic responses all come together to offer a comprehensive and personalized therapy experience. The diagram emphasizes the importance of continuous tracking and personalized recommendations to ensure the system improves over time and meets the user's mental health needs effectively.

## 7.5 Result Analysis

A qualitative analysis of therapists' experiences with online therapy found that online therapy can be a useful addition to traditional therapy, and a dataset is available (https://github.com/cdodiya/Mental-Health-Prediction-using-Machine-Learning-Algorithms/blob/main/survey.csv). The analysis identified three themes shown in Figure 7.5.

The distribution of responses per question for virtual therapy involves analyzing the variety of answers participants provide to survey or feedback questions about virtual therapy experiences, as shown in Figure 7.6. This data can reveal trends, preferences, and areas for improvement in virtual therapy services.

Figure 7.7 shows the number of responses by topic in virtual therapy, which refers to the count of feedback or survey responses categorized under

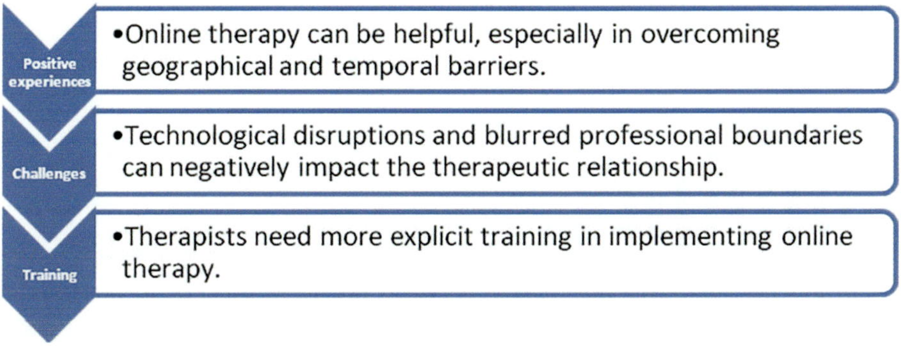


**Figure 7.5** Virtual therapy analysis.

specific topics relevant to virtual therapy. This analysis helps identify which aspects of virtual therapy are most discussed or evaluated, highlighting key areas of interest, concern, or satisfaction.

Figure 7.8 shows the distribution of question and response length in virtual therapy that examines the lengths of survey questions and responses, providing insights into their complexity, engagement levels, and overall participant effort.

Figure 7.9 is a histogram with a density plot overlay, showing the distribution of age data. The $x$-axis represents age, ranging from 0 to 50, while the $y$-axis represents density. The histogram bars indicate the frequency of

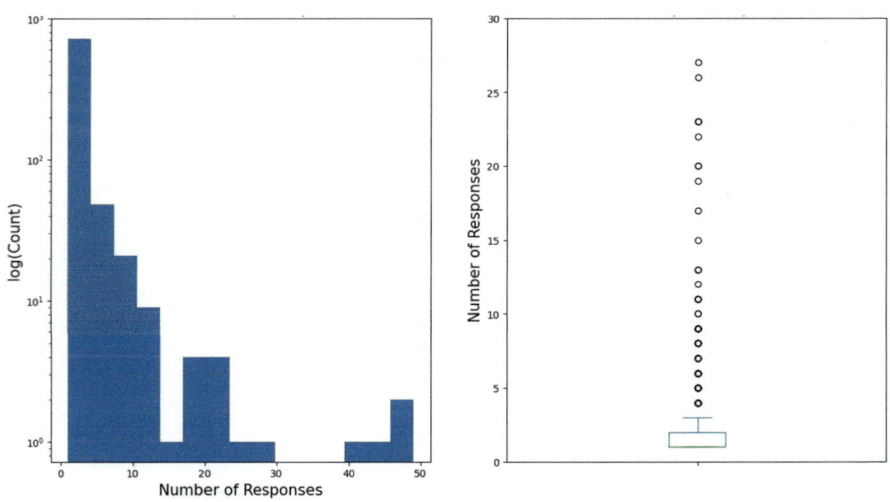


**Figure 7.6** Distribution of responses per question.

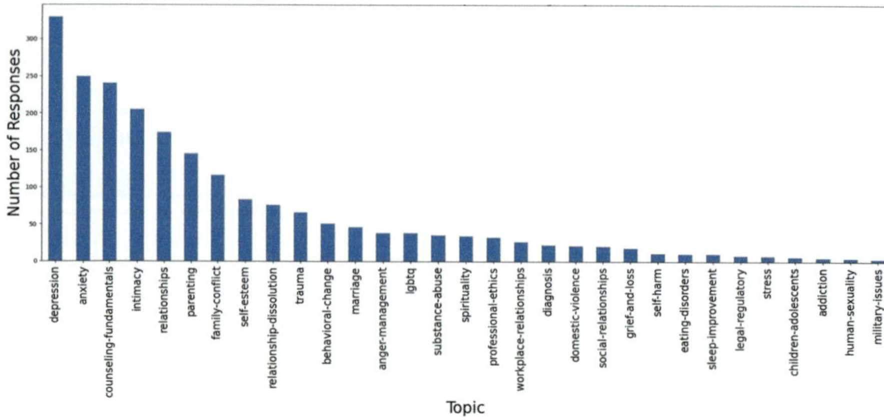


**Figure 7.7** Number of responses by topic.

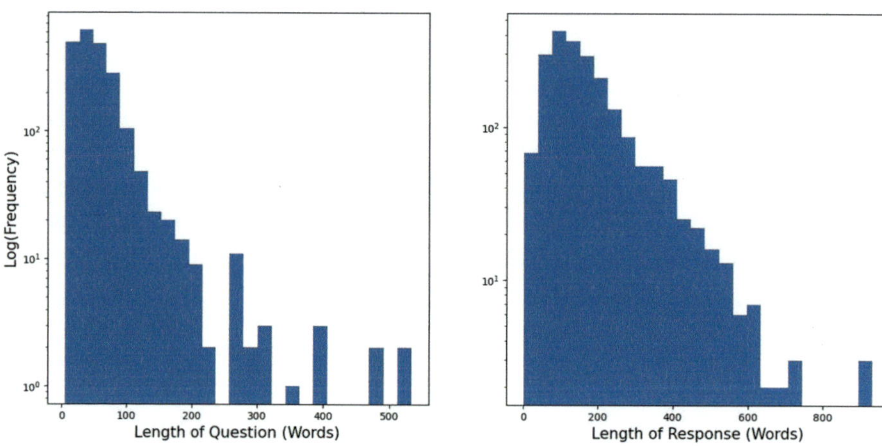


**Figure 7.8** Distribution of question and response length.

different age groups, and the density curve provides a smooth representation of the distribution. The distribution appears to be right-skewed, with most values concentrated between 0 and 20 years, gradually decreasing afterward.

Figure 7.10 shows the total distribution of individuals who received treatment, categorized by gender (female and male). The $x$-axis represents the treatment categories (female and male), while the $y$-axis represents the count of individuals in each category. Both bars are nearly equal in height, indicating that the number of females and males in the dataset is roughly the same, with males slightly outnumbering females.

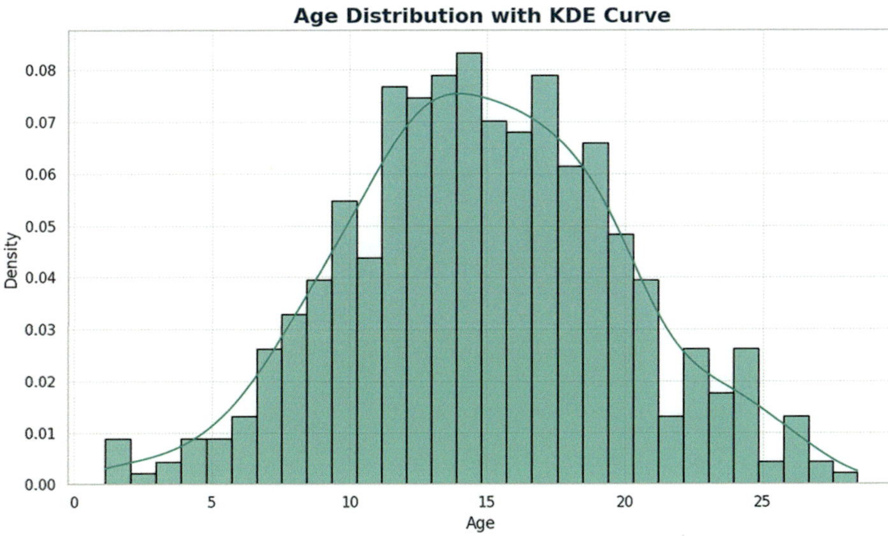


**Figure 7.9** Distribution and density by age.

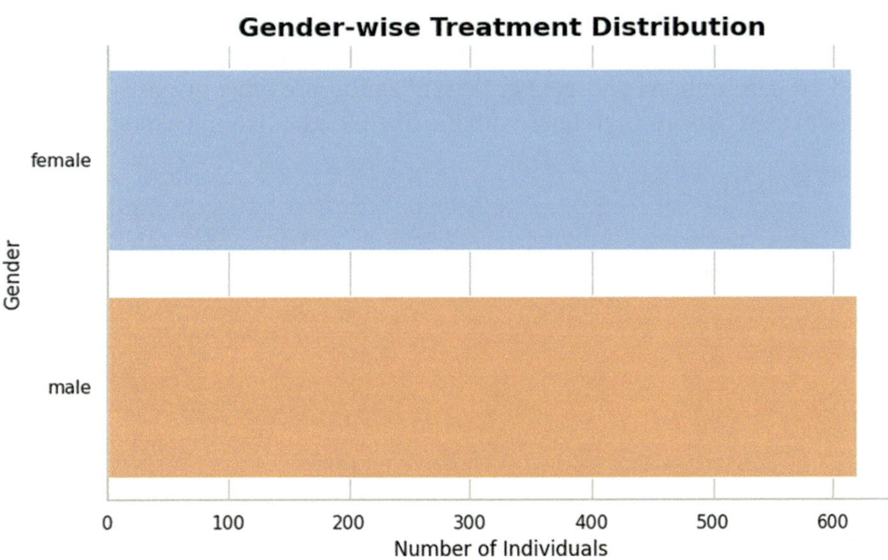


**Figure 7.10** Total distribution by treated or not.

Figure 7.11 (part a) illustrates the probability of experiencing a mental health condition across various age groups and genders. Visualization provides insights into how mental health conditions are distributed among

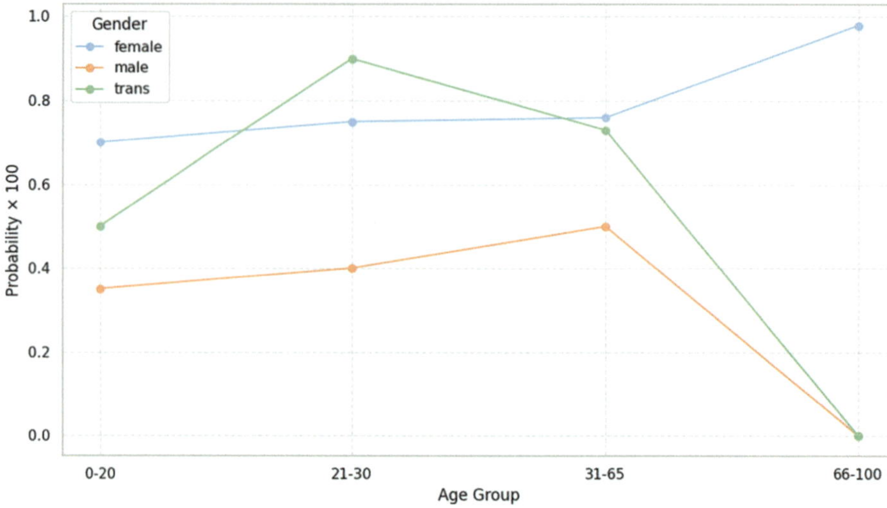


**Figure 7.11** (a) Probability of mental health condition by age and gender.

different demographics, highlighting gender and age-related trends. Figure 7.11 (part b) illustrates the probability of having a mental health condition based on care options and gender. Visualization provides insights into how access to care options correlates with the likelihood of having a mental health

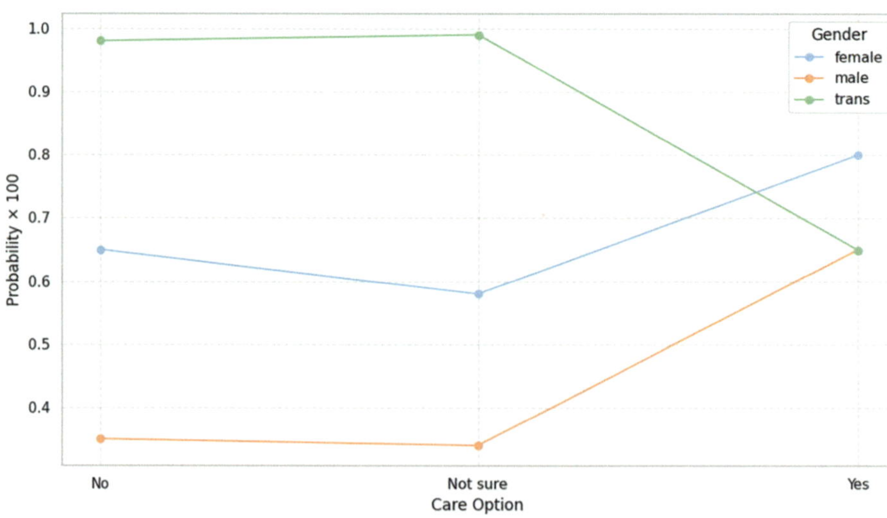


**Figure 7.11** (b) Probability of mental health condition by care and gender.

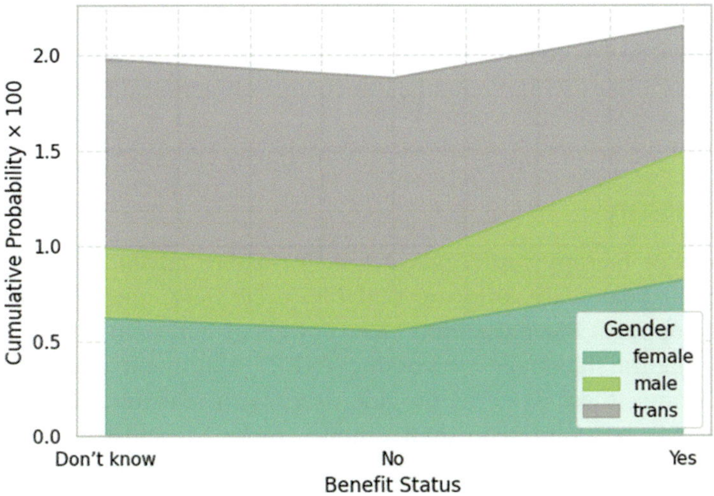


**Figure 7.11** (c) Probability of mental health condition by benefit status.

condition across different gender groups. Figure 7.11 (part c) illustrates the probability of having a mental health condition based on the availability of benefits and gender. This visualization helps illustrate how awareness or availability of benefits correlates with mental health conditions across different gender groups. Figure 7.11 (part d) illustrates the probability of having

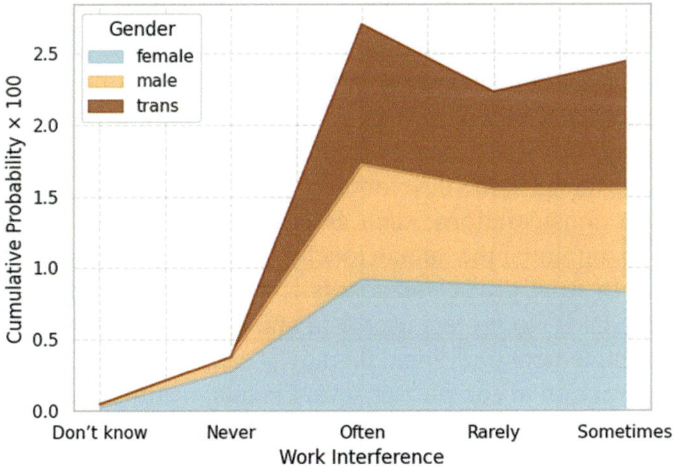


**Figure 7.11** (d) Probability of mental health condition by work interference.

a mental health condition based on how often work interferes with mental health, categorized by gender. This visualization highlights how workplace interference correlates with mental health conditions across different gender groups.

## 7.6 Conclusion

AI-based virtual therapy represents a transformative advancement in the field of mental healthcare, offering scalable, accessible, and personalized therapeutic support. By harnessing the power of artificial intelligence, particularly natural language processing (NLP), sentiment analysis, and machine learning, these systems provide users with real-time, on-demand therapy that can be tailored to their unique emotional and psychological needs. Whether through text or voice interactions, AI-driven virtual therapists can simulate human-like conversations, delivering evidence-based interventions like Cognitive Behavioral Therapy (CBT), mindfulness practices, and emotional support in a way that is both engaging and effective. One of the most significant benefits of AI-based virtual therapy is its ability to overcome traditional barriers in mental healthcare, such as stigma, cost, and limited access to trained professionals. The system's ability to provide therapy at any time, from anywhere, makes it a powerful tool for individuals who may not otherwise have access to in-person therapy due to geographic, financial, or social constraints. Furthermore, by offering a private, non-judgmental environment, virtual therapy helps reduce the stigma that often prevents individuals from seeking help. Additionally, the ability of AI systems to track user behavior, mood, and progress over time allows for continuous adaptation of therapeutic interventions, ensuring that each user receives a personalized experience that evolves with their needs. By collecting and analyzing behavioral data, virtual therapy platforms can refine their responses, improving the system's accuracy and effectiveness with each interaction. Despite these advantages, the development of AI-based virtual therapy also raises important considerations, such as ensuring data privacy, ethical AI practices, and maintaining the human touch in therapy. It is crucial that these systems adhere to strict privacy standards and healthcare regulations (such as HIPAA and GDPR) to protect user data and maintain trust. Additionally, while AI can complement traditional therapy, it cannot fully replace the need for human intervention in complex or severe mental health cases, underscoring the importance of a hybrid approach to mental healthcare. In conclusion, AI-based virtual therapy holds immense potential to revolutionize mental healthcare by making therapy more accessible, affordable, and personalized.

As technology continues to evolve, it promises to be a valuable tool in supporting the mental well-being of individuals around the world, particularly in regions and communities that face significant barriers to traditional therapeutic care. The future of mental health treatment is likely to be increasingly shaped by AI, providing a new pathway for millions to receive the support they need.

## Acknowledgements

I extend my sincere gratitude to Graphic Era (Deemed to be University) for providing the resources and support necessary for this research. My deepest appreciation goes to my mentor, Dr. Sushil Chandra Dimri, whose invaluable guidance and expertise have been instrumental in shaping this study on AI-based psychological treatments for mental health.

I would also like to acknowledge Mr. Bhawnesh Kumar, and Aditya Joshi for their collaborative efforts and insightful discussions, which significantly contributed to the development of this work.

Finally, I am grateful to my family and friends for their unwavering encouragement and motivation throughout this journey. Their support has been a cornerstone of my academic and professional growth.

## References

[1] Jayousi, S., Barchielli, C., Alaimo, M., Caputo, S., Paffetti, M., Zoppi, P., & Mucchi, L. (2024). **ICT in nursing and patient healthcare management: Scoping review and case studies**. Sensors (Basel), 24(10), 3129. https://doi.org/10.3390/s24103129.

[2] Sittig, D. F., Wright, A., Coiera, E., Magrabi, F., Ratwani, R., Bates, D. W., & Singh, H. (2020). **Current challenges in health information technology-related patient safety**. Health Informatics Journal, 26(1), 181-189. https://doi.org/10.1177/1460458218814893.

[3] Riddle, E. W., Kewalramani, D., Narayan, M., & Jones, D. B. (2024). **Surgical simulation: Virtual reality to artificial intelligence**. Current Problems in Surgery, 61(11), 101625. https://doi.org/10.1016/j.cpsurg.2024.101625

[4] Iqbal, A. I., Aamir, A., Hammad, A., Hafsa, H., Basit, A., Oduoye, M. O., Anis, M. W., Ahmed, S., Younus, M. I., & Jabeen, S. (2024). **Immersive technologies in healthcare: An in-depth exploration of virtual reality and augmented reality in enhancing patient care, medical education, and training paradigms**. Journal of Primary

Care & Community Health, 15, 21501319241293311. https://doi.org/10.1177/21501319241293311.

[5] Manzano-Monfort, G., Paluzie, G., & Díaz-Gegúndez, M. et al. (2023). **Usability of a mobile application for health professionals in home care services: A user-centered approach**. Scientific Reports, 13, 2607. https://doi.org/10.1038/s41598-023-29640-7.

[6] Ventola, C. L. (2014). **Mobile devices and apps for health care professionals: Uses and benefits**. P & T, 39(5), 356-364.

[7] González Bermúdez, A., Carramiñana, D., Bernardos, A. M., Bergesio, L., & Besada, J. A. (2024). **A fusion architecture to deliver multipurpose mobile health services**. Computers in Biology and Medicine, 173, 108344. https://doi.org/10.1016/j.compbiomed.2024.108344.

[8] Senbekov, M., Saliev, T., Bukeyeva, Z., Almabayeva, A., Zhanaliyeva, M., Aitenova, N., Toishibekov, Y., & Fakhradiyev, I. (2020). **The recent progress and applications of digital technologies in healthcare: A review**. International Journal of Telemedicine and Applications, 2020, 8830200. https://doi.org/10.1155/2020/8830200.

[9] Guo, J., Huang, X., & Dou, L. et al. (2022). **Aging and aging-related diseases: From molecular mechanisms to interventions and treatments**. Signal Transduction and Targeted Therapy, 7, 391. https://doi.org/10.1038/s41392-022-01251-0.

[10] Sun, X., & Li, X. (2023). **Editorial: Aging and chronic disease: Public health challenge and education reform**. Frontiers in Public Health, 11, 1175898. https://doi.org/10.3389/fpubh.2023.1175898.

[11] Li, X., He, Y., Wang, D., & Rezaei, M. J. (2024). Stroke rehabilitation: From diagnosis to therapy. Frontiers in Neurology, 15. https://doi.org/10.3389/fneur.2024.1402729.

[12] Young, B. M., et al. (2023). Rehabilitation therapy doses are low after stroke and predicted by clinical factors. Stroke, 54(3), 831-839. https://doi.org/10.1161/STROKEAHA.122.041098.

[13] Mendis, S. (2013). Stroke disability and rehabilitation of stroke: World Health Organization perspective. International Journal of Stroke, 8(1), 3-4.

[14] Borghese, N. A., Mainetti, R., Pirovano, M., & Lanzi, P. L. (2013, May). An intelligent game engine for the at-home rehabilitation of stroke patients. Proceedings of IEEE 2nd International Conference on Serious Games and Applications for Health (SeGAH), 1-8.

[15] Jesus, T. S., & Silva, I. L. (2015). Toward an evidence-based patient-provider communication in rehabilitation: Linking communication elements to better rehabilitation outcomes. Clinical Rehabilitation, 30(4), 216-254.

[16] Jani, A. B., Bagree, R., & Roy, A. K. (2017, November). Design of a low-power low-cost ECG EMG sensor for wearable biometric and medical application. Proceedings of IEEE Sensors, 1-3.

[17] Soleimani, M., Ghazisaeedi, M., & Heydari, S. (2024). The efficacy of virtual reality for upper limb rehabilitation in stroke patients: A systematic review and meta-analysis. BMC Medical Informatics and Decision Making, 24, 135. https://doi.org/10.1186/s12911-024-02534-y.

[18] Wu, J., Zeng, A., Chen, Z., Wei, Y., Huang, K., Chen, J., & Ren, Z. (2021). Effects of virtual reality training on upper limb function and balance in stroke patients: Systematic review and meta-meta-analysis. Journal of Medical Internet Research, 23(10), e31051. https://doi.org/10.2196/31051.

[19] Juan, M. C., Elexpuru, J., Dias, P., et al. (2023). Immersive virtual reality for upper limb rehabilitation: Comparing hand and controller interaction. Virtual Reality, 27, 1157–1171. https://doi.org/10.1007/s10055-022-00722-7.

[20] Malvade, P. S., Joshi, A. K., & Madhe, S. P. (2017, April). IoT-based monitoring of foot pressure using FSR sensor. Proceedings of International Conference on Communication and Signal Processing (ICCSP), 635-639.

[21] Pradhan, B., Bhattacharyya, S., & Pal, K. (2021). IoT-based applications in healthcare devices. Journal of Healthcare Engineering. https://doi.org/10.1155/2021/6632599.

[22] Huang, H., Li, X., Liu, S., Hu, S., & Sun, Y. (2018). TriboMotion: A self-powered triboelectric motion sensor in wearable Internet of Things for human activity recognition and energy harvesting. IEEE Internet of Things Journal, 5(6), 4441-4453.

[23] Jani, A. B., Bagree, R., & Roy, A. K. (2017, November). Design of a low-power low-cost ECG EMG sensor for wearable biometric and medical application. Proceedings of IEEE Sensors, 1-3.

[24] Tang, Y. M., Chau, K. Y., Kwok, A. P. K., Zhu, T., & Ma, X. (2022). A systematic review of immersive technology applications for medical practice and education—Trends, application areas, recipients, teaching contents, evaluation methods, and performance. Educational Research Review, 35, 100429. https://doi.org/10.1016/j.edurev.2021.100429.

[25] Kumar, M., Kumar, A., Verma, S., Bhattacharya, P., Ghimire, D., Kim, S.-H., & Hosen, A. S. M. S. (2023). Healthcare Internet of Things (H-IoT): Current trends, future prospects, applications, challenges, and security issues. Electronics, 12, 2050. https://doi.org/10.3390/electronics12092050.

[26] Bonanno, M., De Nunzio, A. M., Quartarone, A., Militi, A., Petralito, F., & Calabrò, R. S. (2023). Gait analysis in neurorehabilitation: From research to clinical practice. Bioengineering, 10, 785. https://doi.org/10.3390/bioengineering10070785.

[27] XR Health. (2024, February 23). Virtual reality rehabilitation: What it is and how it works. XR Health Blog. Retrieved from https://www.xr.health/blog/virtual-reality-rehabilitation.

[28] Institute of Medicine (US) Committee on Disability in America. (2007). The future of disability in America. National Academies Press (US). Retrieved from https://www.ncbi.nlm.nih.gov/books/NBK64952/

[29] Shih, C.-H., Lin, P.-J., Chen, Y.-L., Chen, Y.-L., & Chen, S.-L. (2024). A post-stroke rehabilitation system with compensatory movement detection using virtual reality and electroencephalogram technologies. IEEE Access, PP(99), 1-1. https://doi.org/10.1109/ACCESS.2024.3392513

[30] Bartnicka, J., Herrera, C., Michnik, R., & Garrido, D. (2019). The role of virtual reality and biomechanical technologies in stroke rehabilitation. In Advances in Human Factors in Training, Education, and Learning Sciences (Advances in Intelligent Systems and Computing). https://doi.org/10.1007/978-3-319-93882-0_34

[31] Khan, A., Imam, Y. Z., Muneer, M., et al. (2024). Virtual reality in stroke recovery: A meta-review of systematic reviews. Bioelectronic Medicine, 10, 23. https://doi.org/10.1186/s42234-024-00150-9.

[32] Wang, Y.-L., Wu, W.-X., Yang, C.-C., Huang, S.-M., Chang, C.-C., Li, C.-R., Chiang, S.-L., & Chen, Y.-J. (2024). Heart rate variability biofeedback enhances cognitive, motor, psychological, and autonomic functions in post-stroke rehabilitation. International Journal of Psychophysiology, 203, 112411. https://doi.org/10.1016/j.ijpsycho.2024.112411.

[33] Plisiecki, H., & Sobieszek, A. (2024). Emotion topology: Extracting fundamental components of emotions from text using word embeddings. Frontiers in Psychology, 15. https://doi.org/10.3389/fpsyg.2024.1401084.

[34] Bălan, O., Moise, G., Moldoveanu, A., Leordeanu, M., & Moldoveanu, F. (2019). Fear level classification based on emotional dimensions and machine learning techniques. Sensors, 19(7), 1738. https://doi.org/10.3390/s19071738.

[35] Drigas, A., & Sideraki, A. (2024). Brain neuroplasticity leveraging virtual reality and brain–computer interface technologies. Sensors, 24, 5725. https://doi.org/10.3390/s24175725.

# 8

# Intelligent Blockchain Architectures: Design, Challenges, and Innovations

**Renu Mishra[1], Mamta Narwaria[2], Sandeep Saxena[3] and Murari Kumar Singh[4]**

[1,2]Department of CSE, School of Computing Science and Engineering, Sharda University, India
[3]School of Computer Science and Engineering, IILM University, Greater Noida, India
[4]Department of CSE, GNIOT, Greater Noida, India

**Email:** renutrivedi11@gmail.com; mamta2410@gmail.com; sandeep.research29@gmail.com; mksinghjamia@gmail.com

## Abstract

The chapter explores the convergence of artificial intelligence and blockchain technology, outlining the evolutionary trajectory from conventional blockchain systems to intelligent blockchain architectures. It starts by reviewing the original uses of blockchain and the shortcomings inherent in them, then investigates how AI has transformed the functionality and performance of blockchain. The chapter outlines critical evolution at the crossroads of artificial intelligence and blockchain and includes a holistic comparative study of intelligent and conventional blockchain architectures. A detailed breakdown of the stratified architecture that defines smart blockchains and explores how AI-driven consensus algorithms are revolutionizing blockchain efficiency, security, and scalability is described well. Real-world examples of these technologies in practice through case studies and implementation are presented for a variety of industries. The chapter is concluded by summarizing current challenges and mapping out some of the most promising future research directions in this fast-emerging field, with particular emphasis on architectural innovations that will continue to boost blockchain intelligence and flexibility.



DOI: 10.1201/9788743812739-8

## 8.1 Introduction

Blockchain technology has come a long way from its initial days as a platform for cryptocurrencies to intelligent blockchains integrated with advanced AI. Traditionally, the architecture of blockchains has depended on static consensus protocols and established rules, which tend to suffer from scalability, security, and efficiency problems [1, 2]. Whereas smart blockchains utilize artificial intelligence (AI), machine learning (ML), and automation not only for transaction verification but also for fraud identification, smart contract automation, and network optimization [3]. These systems adapt and learn their operations dynamically from real-time data to provide increased security, scalability, and efficiency. Endowing smart blockchains with AI-powered decision-making will enable autonomous, decentralized, and very efficacious operation, setting them on the next step in the evolution of blockchain and ensuring their relevance in many industries in the near future for a digital transformation.

In earlier days, the traditional architecture of blockchains depended on static consensus protocols and well-defined rules to ensure scalability, security, and efficiency [1, 2]. Smart blockchains go beyond those earlier ones by utilizing AI, ML, and automation for verifying transactions, detecting fraud, automating smart contracts, and optimizing networks [3]. These systems learn and adapt their operations according to real-time data dynamically to provide improved security, scalability, and efficiency. With AI-fueled decision-making, smart blockchains would also operate independently, decentralized, and very effectively, thus marking the next evolution of blockchains and very much contributing to the future of digital transformation across various industries.

The blockchain technology has brought a paradigm shift in the way of recording, securely, and verifying the transactions digitally over peer-to-peer networks. Yet common blockchain architectures have plenty of drawbacks, like constraints on scalability, energy consumption, security, and inefficiencies of smart contract execution. However, emerging technologies like artificial intelligence (AI) and machine learning (ML) can help augment the capabilities of Blockchain, leading to the birth of intelligent blockchains. In this chapter, we discuss the evolution of blockchain along with AI and ML, various developments, integration mechanisms, and future directions.

### 8.1.1 Traditional blockchain systems and their limitations

The concept of blockchain originated with Bitcoin in 2008, establishing a decentralized, immutable ledger system. Not long after, different flavors of blockchains were developed: Ethereum-enabled smart contract functionality and enterprise-grade permissioned blockchains. However, conventional blockchain systems suffer from inherent limitations:

- **Scalability constraints:** Low transaction throughput and high latency related to consensus protocols such as proof-of-work (PoW).
- **Energy inefficiency:** PoW-based mining requires too much computational power, rendering the blockchain network environmentally unsustainable.
- **Security risks:** Susceptibility to 51% attacks, Sybil attacks, and smart contract vulnerabilities.
- **Lack of adaptability:** The deterministic nature of blockchains means they operate on predetermined strategies and do not adapt in real time to changing circumstances.

Despite the numerous benefits of blockchain technology, traditional blockchain architectures suffer from several limitations to their adoption and scalability. Due to such limitations, the concept of Intelligent Blockchains came into existence, which gets rid of the existing technologies' pain points through the infusion of AI and advanced computing techniques.

### 8.1.2 AI and ML integration in blockchain

AI, ML, and automation are used by intelligent blockchains to enhance decision-making, scalability, security, and efficiency, as depicted in Figure 8.1. These architectures are self-optimizing, unlike traditional models, and can facilitate real-time governance, fraud detection, and interoperability [4, 5].

- **AI-powered smart contracts:** Traditional smart contracts execute predefined logic without any flexibility or adaptability. AI-enhanced smart contracts introduce self-learning capabilities where ML algorithms analyze historical transactions to optimize contract execution. AI predicts potential security threats and automatically updates contract conditions, and NLP (Natural Language Processing) enables smart contracts to understand and execute agreements written in human language [7].
- **Enhanced consensus mechanisms:** Consensus mechanisms are at the core of blockchain functionality. AI and ML improve them as AI-driven either proof-of-stake (PoS) or proof-of-work (PoW) to predict node trustworthiness, ensuring fair staking [8, 12]. ML algorithms optimize mining difficulty, reducing energy consumption [9, 10]. AI enables

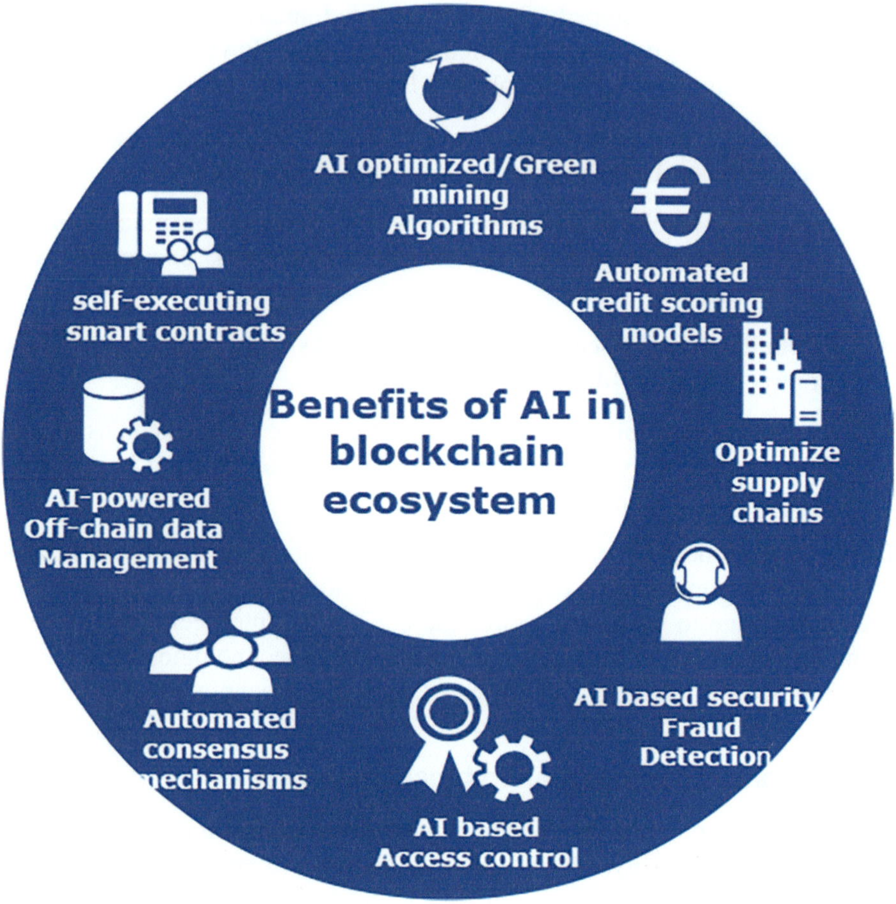


**Figure 8.1** Benefits of AI in blockchain ecosystem.

federated learning for consensus among nodes to enhance consensus efficiency.

- **Security and fraud detection:** Blockchain transactions are immutable, but they are not immune to fraud. AI-based security systems enhance blockchain security through Anomaly Detection by analyzing transaction patterns to detect suspicious activities in real time [11]. ML also predicts user behavior, identifying potential fraud attempts before they occur [13, 14]. Even AI assists in developing security protocols resistant to quantum computing threats using quantum-resistant cryptography [15].

- **Blockchain optimization and scalability:** AI enhances blockchain performance through predictive analytics, which forecasts network congestion and optimizes transaction throughput. Machine learning-based data compression techniques reduce storage requirements, while AI dynamically distributes workloads across multiple blockchain shards using sharding and AI-based load balancing, thereby improving overall efficiency [17].
- **Interoperability and cross-chain communication:** AI facilitates seamless communication between different blockchain networks by automating cross-chain transactions through AI-driven oracles that enable secure asset transfers, while ML-based interoperability protocols optimize data translation and validation across heterogeneous blockchain platforms [18].

### 8.1.3 Transformative stages in AI-blockchain collaboration

This evolving collaboration between AI and blockchain is illustrated in Figure 8.2, which outlines the progression through three key phases, ultimately leading to the development of autonomous, trustless ecosystems where intelligent systems can operate securely and transparently with minimal human oversight.

#### Phase 1: AI's role in early blockchain innovations (2015–2018)

During this phase, blockchain-based AI applications for data authentication, predictive analytics, and smart contract auditing have been developed [17]. AI smart automation took its first steps to be integrated into blockchain networks as well, mainly focused on market predictions or fraud recognition. AI models were used to study blockchain transactions and identify anomalies and enhance security [11]. However, the actual blockchain architecture has not built extensive integration with AI.

#### Phase 2: AI-driven blockchain solutions (2019–2022)

AI also started to optimize internal blockchain processes, focusing on better consensus mechanisms and scalability. During this period, decentralized AI marketplaces (SingularityNET) emerged, allowing artificial intelligence models to be exchanged securely across blockchain networks [19]. AI-driven mining optimization, smart contract validation, and fraud detection contributed to a boost in performance for both decentralized applications (DApps) and enterprise blockchains [20].

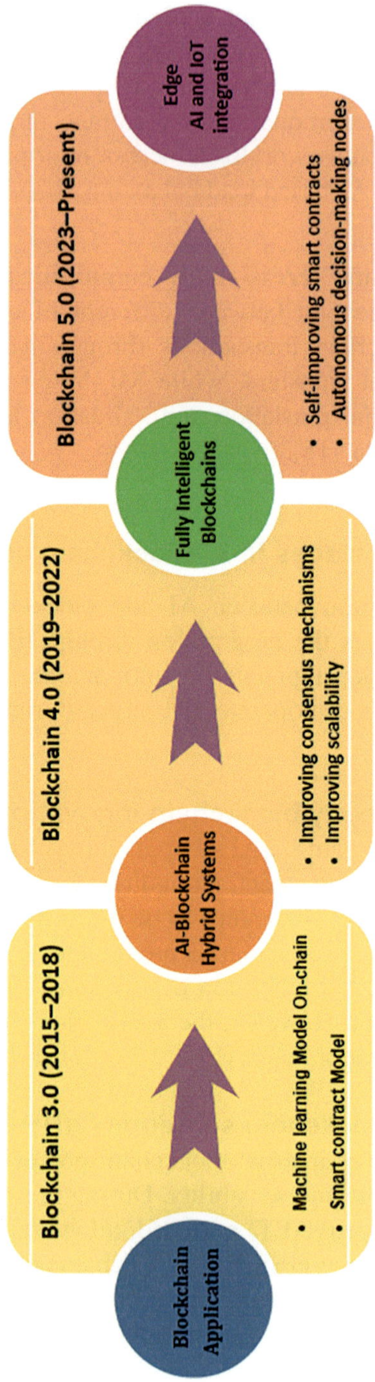


**Figure 8.2** Evolutionary phases of AI-blockchain integration.

**Phase 3: Fully intelligent blockchains (2023–present)**
Blockchain topologies now include extensive AI so that smart contracts will be more self-improving and independent decision-making nodes with a life of their own [21]. A decentralized decision-making process is further complemented by AI governance models, and Edge AI, together with connectivity of things, make real-time decentralized intelligence possible [22, 23]. To build fully autonomous intelligent and computer networks, newcomers in blockchain reinforce AI for not only cryptocurrency compliance rules but also adaptive consensus, precise security analytics that outpace traditional methods based on historic data for an accuracy score.

## 8.2 Comparative Analysis of Intelligent and Traditional Blockchain Architectures

These intelligent blockchain architectures integrate the power of artificial intelligence (AI), machine learning, and advanced cryptographic techniques to improve the scalability of decentralized systems, security, and adaptability. AI-based consensus mechanisms like proof-of-learning [24] and AI-enhanced proof-of-stake (PoS) [25] have been investigated to maximize energy utilization and network capabilities. Autonomous anomaly detection [26] algorithms and self-adaptive smart contracts [27] monitor, identify, and neutralize cyberattacks to effectively enhance the efficiency and capacity of the system.

Scalability is still one of the most critical issues; work on sharding [28], DAG-based architecture [29], and Layer 2 rollups [30] has shown a lot of promising solutions. Various interoperability frameworks [31] and cross-chain communication protocols [32] have been proposed to support the smooth exchange of data between disparate blockchain networks. The need for ethical concerns in AI, such as AI bias [33], data privacy risks [34], and algorithmic accountability [35], prioritizes transparent, explainable AI in blockchain governance.

The effects of regulation, from legal uncertainties around smart contract enforcement [36] and decentralized governance models [37], securities laws [38] to tax policy [39], must be considered, along with the challenges of raising funds [40] and the need for regulatory sandboxes [41], while co-creating best practices alongside the regulatory environment [42].

According to recent work on quantum-resistant cryptography [43] and federated learning for blockchain security [44] shows promising pathways for secure and privacy-preserving applications. Artificial intelligence for enhanced compliance and security in next-generation blockchain ecosystems. The adoption of AI-assisted compliance mechanisms [45] and automated cybersecurity frameworks [46] is bolstering the next-era blockchain environments.

**Table 8.1** Key differences between traditional and intelligent blockchain architectures.

| Feature | Conventional blockchain | Intelligent blockchain |
|---|---|---|
| **Smart contracts** | Static, rule-based execution | Self-learning, adaptive, and AI-optimized |
| **Consensus mechanism** | PoW, PoS (energy-intensive, slower) | AI-driven PoS, adaptive PoW, federated learning |
| **Security and fraud Detection** | Manual or predefined rule-based detection | AI-powered real-time anomaly detection and prevention |
| **Scalability** | Limited by network congestion and block size | AI-based sharding, predictive load balancing |
| **Data processing** | Sequential, requires high storage | AI-based data compression & optimization |
| **Interoperability** | Limited cross-chain communication | AI-driven interoperability and automated oracles |
| **Network efficiency** | High latency due to complex computations | Predictive analytics for optimized transactions |

As these technologies continue to develop, smart blockchain architectures will redefine trust, efficiency, and automation in decentralized networks.

The incorporation of AI into blockchain technology brings several improvements that distinguish Intelligent Blockchains from Traditional Blockchain Architectures. Table 8.1 represents a comparison based on key characteristics.

Intelligent blockchain improves scalability, security, and efficiency by leveraging AI. They bring in self-learning contracts, AI-optimized consensus, and automated fraud detection. They offer real-time decision-making, interoperability, and adaptive governance as opposed to traditional blockchains.

## 8.3 Methodology of Intelligent Blockchains

The proposed approach is a multi-layered integration of artificial intelligence (AI) and other decision-making modules to ensure adaptability, security, and scalability. The following outlines the key methodological components:

### 8.3.1 Layered architecture of intelligent blockchains

By combining decentralized principles with intelligent automation, adaptive consensus, predictive analytics, and AI-driven security, intelligent

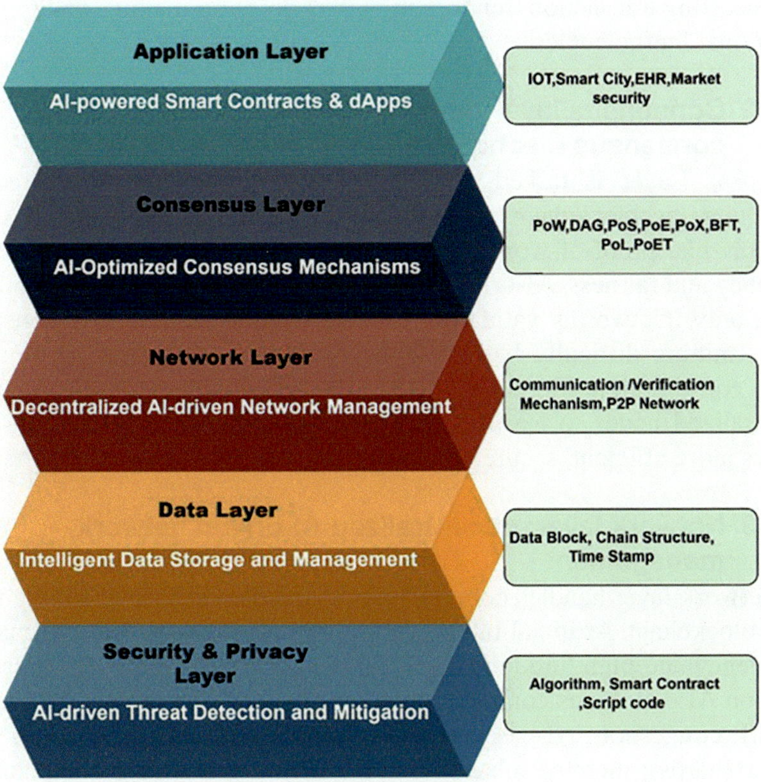


**Figure 8.3** Layered architecture of intelligent blockchains.

blockchains are set to revolutionize the future of decentralized ecosystems. Figure 8.3 illustrates the proposed layered architecture that encapsulates the synergy between blockchain infrastructure and AI functionalities across various system layers.

This layered architecture integrates AI at multiple levels to enhance the functionality, efficiency, security, and scalability of blockchain networks.

#### 8.3.1.1 Application layer: AI-powered smart contracts & dApps

The application layer is the topmost layer where AI-powered smart contracts and decentralized applications (dApps) operate. Unlike traditional smart contracts that execute static, predefined logic, AI-enhanced smart contracts learn and adapt based on historical transaction data, optimizing execution and preventing security risks. AI also improves natural language processing (NLP)-based contracts, allowing automated agreement interpretation. Additionally, AI-powered predictive analytics enhances dApp performance by forecasting transaction trends, automating decision-making, and reducing operational inefficiencies.

#### 8.3.1.2 Consensus layer: AI-optimized consensus mechanisms

Consensus Layer is tasked with transaction verification and agreement throughout the blockchain network. AI improves conventional consensus protocols like proof-of-work (PoW) and proof-of-stake (PoS) to maximize efficiency and fairness. AI-driven PoS constantly assesses node credibility to ensure only trustworthy validators are employed. Adaptive PoW constantly adjusts mining difficulty based on the network environment to minimize energy consumption. Moreover, federated learning-based consensus enables decentralized nodes to learn distributed data, which makes the consensus process more efficient, secure, and adaptive.

#### 8.3.1.3 Network layer: decentralized AI-driven network management

The network layer handles communication and data exchange among nodes of the blockchain. Artificial Intelligence optimizes network routing, enabling low latency and high throughput for transaction processing. Load balancing based on AI distributes computation tasks efficiently and therefore prevents network congestion. AI-based sharding divides the blockchain into small chains (shards), thereby enhancing scalability and reducing computational overhead [49]. This layer also enables self-healing networks, where AI automatically identifies inefficiencies, reroutes traffic, and fixes connectivity issues in real-time.

#### 8.3.1.4 Smart data storage and management layer

The data layer provides effective storage, retrieval, and management of blockchain data. AI-driven compression algorithms minimize storage needs, supporting quicker transactions without sacrificing security. AI renders off-chain and storage methods more scalable and economical without overwhelming the blockchain. AI-driven predictive caching algorithms improve data access speeds through predicting the most frequently accessed records, making retrieval more effective and query response quicker. These technologies solve the scalability and storage problems of traditional blockchain systems.

#### 8.3.1.5 Security and privacy layer

AI-Driven Threat Detection and Mitigation Security & Privacy Layer defends the blockchain from cyber-attacks, fraud, and data breaches. AI improves real-time anomaly detection, which detects unusual activity like fraudulent transactions and Sybil attacks. AI-based threat intelligence systems regularly scan the blockchain network for potential weaknesses and adjust security in response. Quantum-resistant cryptographic algorithms are also being created with AI to defend blockchain networks from future quantum computing attacks [50]. AI also makes privacy-preserving mechanisms like federated learning and homomorphic encryption possible to securely process data without breaching confidentiality.

### 8.3.2 AI-driven consensus mechanisms in blockchain

AI-consensus mechanisms utilize artificial intelligence to enhance the blockchain verification process and achieve much greater levels of efficiency and resource utilization than conventional methods.

a) *Reinforcement learning for consensus optimization*: RL-based consensus optimization is becoming a fairly new area of research aimed at using reinforcement learning methods toward improving consensus methods for multi-robot systems, distributed networks, and decentralized optimization problems. Consensus optimization becomes a key objective in blockchain networks, federated learning, and swarm robotics control, as well as sensor and IoT networks. Agents agree and converge to a common value or decision, for example, electing a leader or computing an average. It also attempts to minimize consensus error, accelerate convergence, and compromise between efficiency, robustness, and cost of communication. In the reinforcement learning strategy, Agents learn how to modify their states from localized estimates (state), apply update rules (action), enhance group performance (reward), and pursue learned strategies (policy).

b) *AI-enhanced proof-of-work (PoW) and proof-of-stake (PoS)*: AI enhances blockchain consensus algorithms such as proof-of-work (PoW) and proof-of-stake (PoS) through increased efficiency, security, and scalability. In PoW, AI maximizes mining approaches, enhances energy efficiency, identifies fraudulent behavior, and dynamically modifies difficulty levels. In PoS, AI improves validator choice, identifies security risks, maximizes staking strategy, and promotes energy-efficient consensus. In general, integration of AI with blockchain consensus increases energy efficiency, security, scalability, and equity, making the networks more robust and sustainable.

c) *Adaptive byzantine fault tolerance (BFT) using AI:* AI boosts byzantine fault tolerance (BFT) through improved security, efficiency, and scalability in distributed networks. AI-based BFT provides Dynamic trust scoring and anomaly detection for recognizing malicious nodes. Adaptive Consensus Mechanisms that dynamically configure quorum and voting thresholds. Predictive failure analysis and automated recovery to ensure no downtime. Energy-efficient consensus by reducing energy consumption while still maintaining robust operations. AI-powered BFT delivers strong, secure, and scalable fault tolerance in blockchain and decentralized networks.

### 8.3.3 Intelligent smart contracts and autonomous execution

Intelligent smart contracts enhance traditional blockchain-based agreements by incorporating AI and ML capabilities for adaptive decision-making and autonomous execution without human intervention.

a) *Self-learning smart contracts:* Self-learning smart contracts are an advanced evolution of traditional blockchain-based contracts, integrating artificial intelligence (AI) and machine learning (ML) to create adaptive, intelligent agreements. Unlike conventional smart contracts, which execute predefined rules rigidly, self-learning contracts continuously analyze data, identify patterns, and optimize their execution based on real-time interactions. This capability allows them to enhance efficiency, improve security, and reduce costs, making them highly effective for decentralized applications (dApps), finance, supply chain management, and beyond.

   One of the key advantages of self-learning smart contracts is their ability to optimize execution parameters dynamically. For instance, these contracts can adjust gas fees based on network congestion, predict optimal execution times, and refine resource allocation to minimize unnecessary costs. Additionally, they can learn from past transactions and user behavior to enhance decision-making, making them more responsive to changing conditions in the blockchain ecosystem. This level of adaptability is crucial in decentralized finance (DeFi) and automated trading, where market conditions fluctuate rapidly.

   Security is another area where AI-powered smart contracts provide significant improvements. Traditional smart contracts are vulnerable to hacks, fraud, and malicious exploits due to their immutable nature – once deployed, they cannot be modified. However, self-learning contracts can detect anomalies, identify potential threats, and take proactive security measures before an attack occurs. By leveraging AI-driven fraud detection systems, these contracts can recognize suspicious transaction patterns and adjust their security protocols dynamically, reducing the risks associated with smart contract vulnerabilities.

Another crucial benefit is in dispute resolution and error correction. In many blockchain applications, errors or disputes require human intervention, slowing down the process and increasing costs. Self-learning smart contracts can analyze past transactions, identify inconsistencies, and propose automated solutions based on historical data. This capability makes them particularly useful in legal agreements, insurance claims, and supply chain management, where contract terms may need to be adjusted based on unforeseen circumstances.

Overall, self-learning smart contracts represent the next generation of blockchain automation, combining the transparency and security of blockchain with the adaptability and intelligence of AI. By evolving over time, optimizing execution strategies, and proactively mitigating risks, these contracts enhance efficiency, security, and flexibility, paving the way for more advanced decentralized ecosystems and intelligent financial systems.

b) *AI-based anomaly detection in smart contracts:* Smart contracts are automated contracts on a blockchain that run according to pre-established rules. Yet, they are susceptible to numerous anomalies, including fraudulent transactions, security attacks, and unintended execution faults, which can cause financial losses and inefficiencies in the system. AI-driven anomaly detection improves the security and reliability of smart contracts by using machine learning (ML) and artificial intelligence (AI) to detect and prevent unusual patterns, providing a more secure decentralized ecosystem. One of the most significant advantages of anomaly detection powered by AI is fraud avoidance. Smart contracts based on traditional models do not possess the capability to dynamically reallocate or identify malicious activities after deployment. AI algorithms, especially deep learning and pattern detection models, can scrutinize an enormous volume of transaction information and identify unusual behavior like double-attempts at spending, Sybil attacks, or suspicious withdrawals. Through ongoing observation of the blockchain network, AI can identify high-risk transactions and initiate preventive security actions, including suspending contract execution or demanding additional validation processes.

Another important benefit is error detection and prevention. Flaws in code, logic bugs, and unwanted vulnerabilities in smart contracts may be taken advantage of by attackers, resulting in great financial losses. AI systems can utilize static and dynamic code analysis to analyze smart contract logic prior to deployment, detecting flaws that may be exploited. Moreover, post-deployment monitoring enables AI to identify abnormal contract behavior, for instance, untimely fund transfers or deviant execution times, so errors are quickly spotted and resolved. AI also plays a vital part in adaptive security measures. Unlike conventional security audits, static and pre-deployment in nature, AI-powered anomaly detection adapts by constantly learning from in-time blockchain usage. AI achieves this using reinforcement learning and neural networks, thereby dynamically tweaking the security parameters of smart contracts against changing threats. For instance, if an AI system recognizes spikes in anomalous contract interactions, it can reconfigure access controls or introduce supplementary verification procedures to neutralize risk. In addition to security, AI-driven anomaly detection also boosts regulatory compliance in blockchain environments. Decentralized finance (DeFi) platforms and enterprise blockchain solutions often need to comply with financial regulations that necessitate monitoring transactions for illicit activities. By examining smart contract transactions and highlighting unusual behavior, AI-powered models can identify money laundering, insider trading, or regulatory violations. This ensures smart contracts run transparently and in accordance with legal expectations. AI-based anomaly detection is revolutionizing the security and reliability of smart contracts by providing real-time fraud detection, error prevention, and adaptive security measures. By leveraging AI's ability to analyze vast blockchain datasets and detect irregularities, smart contracts can become more resilient, secure, and compliant. As blockchain adoption grows, integrating AI-driven anomaly detection will be essential for protecting decentralized ecosystems from evolving threats and ensuring trust in automated transactions.

### 8.3.4 Integration of IoT and edge computing with intelligent blockchains

The convergence of internet of things (IoT), edge computing, and intelligent blockchains is transforming industries by enabling secure, decentralized, and real-time data processing. Traditional IoT architectures face challenges such as scalability, security risks, latency, and centralized control, which can lead to data breaches, inefficiencies, and high operational costs. By integrating edge computing and blockchain with AI-driven intelligence, IoT systems become more secure, autonomous, and scalable. IoT devices generate vast amounts of data that require secure and tamper-proof storage. Blockchain provides decentralized authentication, encryption, and immutability, preventing cyberattacks, data breaches, and unauthorized access [48]. AI-powered blockchain systems analyze. IoT data in real time, detecting anomalies and ensuring secure device-to-device communication. Example: IBM & Samsung's ADEPT (Autonomous Decentralized Peer-to-Peer Telemetry) – This blockchain-powered IoT platform enables smart devices to communicate and transact autonomously, reducing the need for centralized control. AI-Powered Blockchain for Smart IoT Contracts: AI-driven smart contracts automate and enforce IoT-based transactions, ensuring real-time payments, predictive maintenance, and autonomous device interactions. These contracts self-adjust based on AI analysis of IoT data, optimizing performance while reducing manual intervention. Example: Filament's Blockchain IoT Solutions – Uses blockchain-based smart contracts to automate machine-to-machine (M2M) transactions, enabling IoT devices to independently buy and sell computing resources.

The integration of IoT, edge computing, and intelligent blockchain is revolutionizing digital ecosystems by enabling secure, decentralized, and real-time decision-making. By leveraging AI-driven analytics, blockchain security, and edge computing efficiency, businesses and smart infrastructures can enhance automation, reduce risks, and improve operational scalability.

## 8.4 The Mathematical Architecture

The mathematical models that can be adapted or extended to represent and analyze Intelligent Blockchain Architectures help in quantifying, optimizing, and simulating various components such as consensus, smart contracts, and AI decision-making. There are various base mathematical models that can be explored and employed to create an Intelligent Blockchain Architecture. Some are discussed as follows:

### 8.4.1 Game theoretic models

This is used to analyze incentives, cooperation, and strategic interactions among nodes (especially in consensus and smart contract execution). These models are used to design incentive-compatible consensus protocols, prevent selfish mining or collusion attacks, and analyze Nash equilibria in smart contract markets. The model can be defined as follows.

**Model:** G = ⟨N,A,u⟩, where:

N: Set of agents (nodes, miners, validators)
A: Set of actions (e.g., mining, validating, forking)
ui: A→R: Utility function for each agent *i*

### 8.4.2 Markov decision processes (MDP) for smart contracts

Model decision-making in dynamic environments where smart contracts or agents adapt based on observed states. MDP are significant in designing reinforcement learning-based smart contracts, Dynamic pricing in DeFi protocols. The model can be defined as follows:

**Model:** M = ⟨S,A,P,R,γ⟩, where:

S: Set of states (e.g., contract status, network state)
A: Set of actions (contract triggers, payments, penalties)
P(s′|s,a): Transition probability
R(s,a): Reward function
γ: Discount factor

### 8.4.3 Queuing theory for transaction throughput optimization

Models the system as a set of queues (e.g., mempool, block creation) to analyze delays, throughput, and service rates. AI-powered load balancing in high-traffic blockchain environments, Adaptive gas pricing strategies. The model definition is

**Model:** M/M/1 or M/G/1 queues

λ: Transaction arrival rate
μ: Service rate (block creation rate)
Performance metrics: latency, queue length, utilization $\rho = \lambda/\mu$

### 8.4.4 Graph neural networks (GNNs) for blockchain networks

Used to model relationships and learning tasks over the blockchain's peer-to-peer graph structure. GNNs are generally used for node behavior prediction, Anomaly detection in transaction graphs, and reputational scoring. The model can be depicted as follows:

**Model:** Let G = (V,E) be the network graph

Node features *hv* updated as:

$$h_v^{l+1} = \sigma \sum_{u \in N(v)} W^l h_u^l + b^l \tag{8.1}$$

### 8.4.5 Differential equations for protocol dynamics

This represents demonstrating the evolution of variables (e.g., token supply, consensus parameters) over time. Application includes: Dynamic stability of token economies, protocol tuning with AI control. The model can be written as

**Model:**

$$\frac{dx}{dt}(t) = f\left(x(t), u(t), t\right) \tag{8.2}$$

x(t): System state (e.g., network congestion, staking rates) and
u(t): Control input (e.g., AI-tuned parameters).

### 8.4.6 Bayesian networks for trust & reputation

This model's uncertainty in participant behavior using probabilistic graphical models. These models are used in trust scoring systems, fraud detection via probabilistic inference. It can be represented as follows:

**Model:**

$$P\left(X_1, \ldots. X_n\right) = \prod_{i=1}^{n} P\left(X_i \mid Parents\left(X_i\right)\right) \tag{8.3}$$

Nodes represent agents and edges represent dependencies.

## 8.5 Integration of Federated Learning and Blockchain and Its Mathematical Framework

In conventional machine learning systems, usually data is **centralized** at a single server for data aggregation for model training, so posing major issues of **privacy risks** and **single points of failure**. Federated learning (FL) preserves local data privacy by allowing many clients to jointly train a global model without sharing raw data, providing a distributed alternative. Deployed inside a blockchain system, FL gains even more from improved decentralization, transparency, and tamper resistance. As shown in Figure 8.4, blockchain uses

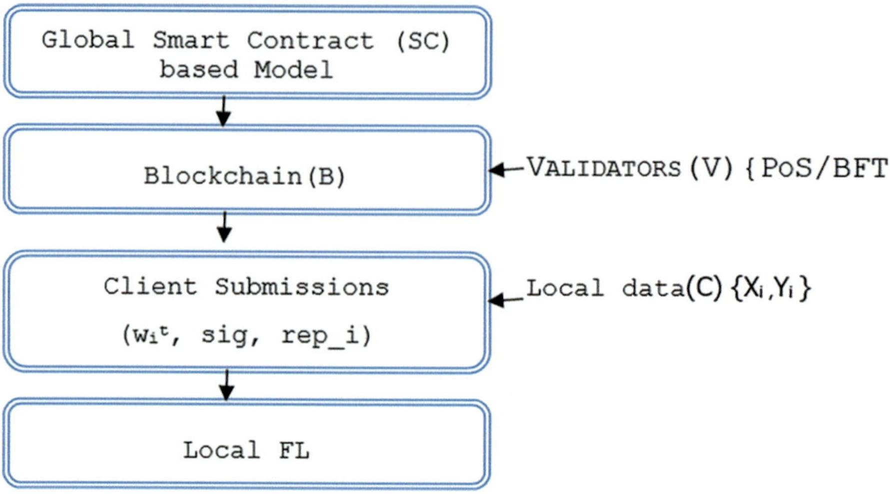


**Figure 8.4** Mathematical framework for the integration of federated learning and blockchain.

smart contracts to automate incentive systems and allows safe, unchangeable recording of model updates, so encouraging consistent client involvement. This integration offers a strong paradigm for distributed environments' scalable, secure, and privacy-preserving machine learning.

where

$C$ = {C11ree CN} Set of federated clients with local data
$B$ = {B11 BT} Sequence of blockchain blocks recording updates
SC = Smart Contract that manages aggregation, validation, and rewards
$V$ = Validators or miners who verify transactions and enforce consensus
$W^t$ = Global model at training round $t$

### 8.5.1 Federated learning model

Each client $C_i$ minimizes its own local objective:

$$f_i(w) = \frac{1}{n_i} \sum_{j=1}^{nj} l\left(w; x_{ij}, y_{ij}\right) \tag{8.4}$$

Global FL objective is done as where

$$\min_{w} F(w) = \sum_{i=1}^{N} p_i f_i(w), \text{ where } p_i = \frac{n_j}{\sum_{j=1}^{N} (n_j)} \tag{8.5}$$

Local update rule is applied as

$$w_i^{t+1} = w_i^t - \eta \nabla f_i\left(w_i^t\right) \tag{8.6}$$

Where:

$F(w)$: global loss function
$f_i(w)$: loss function on client $i$'s local data
$N$: total number of participating clients (devices or nodes)
$p_i$ : proportional weight of client $i$, equal to the number of samples at client $i$ divided by the total number of samples across all clients
$w$: Global model parameters to be learned
$n_i$: number of samples at client $i$
where $\eta$ is the learning rate.
Then clients send model updates to a **global aggregator**.

## 8.5.2 Blockchain interaction

Instead of a central server, model updates are **broadcast to a blockchain network**.

Let $B_t$ represent the **block** at round $t$, containing:

$$B_t = \left\{\left(i, w_i^t sig_i\right) \mid i \in S_t\right\} \tag{8.7}$$

$S_t$: selected client subset
$\text{sig}_i$: digital signature ensuring authenticity
Encrypted or privacy-preserving model update wit, hash, and signature are broadcast

## 8.5.3 Validation and aggregation via smart contract

The are responsible for verifying signatures and updating the format. smart contract logic deployed on the blockchain computes:

$$w_i^{t+1} = \sum_{i \in s_t} p_i \; w_i^t \tag{8.8}$$

This can be embedded in a smart contract, which inevitably validates signatures and verifies update quality (e.g., gradient norm thresholds) followed by aggregating models. Then the distribution of the global model is done as the Aggregated model wt+1 is then broadcast to all clients for every subsequent round.

## 8.6 Real-World Applications and Case Studies

Having covered the theoretical foundations and architectural components of smart blockchain systems, we proceed to their actual deployments in sectors. The following examples illustrate how the convergence of AI, ML, and blockchain technologies is delivering real-world benefits and addressing real-world challenges. These examples not only validate the architectural fundamentals covered above but also provide valuable insights into implementation strategies, performance outcomes, and learnings of deploying smart blockchain systems in production.

### 8.6.1 Financial services (fraud detection, risk assessment)

The integration of intelligent blockchain architectures with artificial intelligence (AI) and machine learning (ML) is transforming financial services by enhancing fraud detection, risk assessment, and overall security. These innovations help financial institutions improve transparency, efficiency, and regulatory compliance while reducing operational risks. Below are key real-world applications and case studies demonstrating the impact of intelligent blockchain in financial services.

#### 8.6.1.1 Case study 1: HSBC's blockchain-based fraud prevention

AI-powered blockchain solutions are used to detect fraudulent activities in financial transactions by analyzing patterns, identifying anomalies, and preventing unauthorized access. Traditional fraud detection methods rely on static rule-based systems, but AI-enhanced blockchain platforms leverage real-time data analytics and predictive modeling to identify suspicious transactions before they occur. HSBC has implemented blockchain solutions integrated with AI to track and authenticate trade finance transactions. By using a distributed ledger, HSBC ensures that financial records are tamper-proof, reducing the risk of double invoicing, identity theft, and synthetic fraud. The system continuously analyzes transaction patterns and flags unusual activities, making fraud detection more effective.

#### 8.6.1.2 Case study 2: bloom decentralized credit scoring

Blockchain systems with AI capabilities improve credit risk assessment by offering a decentralized, transparent, and unchangeable record of a borrower's financial history. Blockchain guarantees safe and reliable financial records, which enhances credit risk assessments. Traditional credit score depends on centralized databases that are subject to manipulation or lack complete data. Using decentralized identity verification and other data sources, Bloom, a blockchain-based credit score platform, incorporates AI to evaluate borrower risk. Bloom offers a more precise and impenetrable credit score by examining a borrower's financial activity both on and off the chain. By enabling underbanked people to obtain loans without depending on conventional credit bureaus, this invention has improved financial inclusion.

#### 8.6.1.3 Case study 3: JPMorgan's AI and blockchain AML system

Blockchain technologies powered by AI are used by financial institutions to automate AML compliance through transaction tracking, suspicious activity detection, and audit trail generation. Blockchain systems with AI enhancements improve the accuracy of identifying illegal transactions, including money laundering and terrorism financing, and decrease false positives. JPMorgan has created a blockchain network driven by AI to track international transactions in real time. By examining transaction flows, user activity, and historical data, this technology finds irregularities in financial transactions. By leveraging blockchain's immutability and AI's pattern recognition, JPMorgan significantly reduces fraudulent transactions and enhances regulatory compliance.

#### 8.6.1.4 Case study 4: AXA's smart contract-based insurance payouts

By automating contract execution based on preset criteria and real-time risk assessments, AI-enhanced smart contracts simplify risk management. Based on AI-driven risk analysis, these smart contracts self-correct to keep financial agreements safe and compliant. Leading insurance provider AXA unveiled fizzy, a blockchain-powered flight insurance platform that uses smart contracts to process claims automatically. AI analyzes flight data, and when a delay is detected, the smart contract automatically triggers payouts, eliminating disputes and speeding up compensation. By integrating AI-driven fraud detection, risk assessment, and automated compliance mechanisms, intelligent blockchain architectures are revolutionizing financial services. Financial institutions that use AI-powered blockchain solutions enjoy increased security, transparency, and efficiency, which lowers fraud risks and improves operational decision-making. As the financial industry develops, AI-enhanced blockchain technology will be crucial in forming the future of safe and intelligent financial ecosystems.

### 8.6.2 Supply chain management (AI-driven blockchain tracking)

By giving supply chain managers real-time visibility into the flow of commodities, AI-driven blockchain tracking improves security, efficiency, and transparency. By utilizing immutable ledgers, smart contracts, and predictive analytics, blockchain solutions driven by artificial intelligence (AI) help address the issues of fraud, inefficiencies, and lack of traceability that plague traditional supply chain systems. The main uses and case studies that illustrate the practical effects of intelligent blockchain architectures in supply chain management are listed below.

#### 8.6.2.1 Case study 1 (IBM food chain): Ensuring food safety with IBM food trust

End-to-end product traceability is made possible by AI-driven blockchain systems, which safely capture supply chain data, guarantee authenticity, and lower the danger of counterfeiting. AI helps companies maintain quality control by analyzing blockchain data to identify irregularities in supply chain transactions. IBM Food Trust tracks food items from farm to table using blockchain and artificial intelligence. This software is used by companies such as Walmart and Nestlé to track the origins of food, identify sources of contamination, and stop fake items from getting into the supply chain. The system may anticipate such disruptions and suggest substitute suppliers by incorporating artificial intelligence.

#### 8.6.2.2 Case study 2: TradeLens & Maersk: Streamlining international shipping

By using real-time blockchain data to enforce supplier agreements, verify deliveries, and initiate payments, AI-enhanced smart contracts simplify logistics procedures. This minimizes disagreements between distributors and suppliers and reduces manual paperwork. Maersk and IBM's TradeLens are working together to modernize international shipping procedures using blockchain and artificial intelligence. By automating supply chain documentation, customs clearance, and container tracking, the system lowers operating expenses by 30% and shipping delays by 40%. Shipping routes are optimized using AI-driven analytics according to environmental parameters and demand trends.

#### 8.6.2.3 Case study 3: Blockchain and AI-powered inventory management at Amazon

AI forecasts inventory demand, cuts waste, and avoids stock shortages by evaluating real-time supply chain data. Blockchain minimizes inconsistencies by ensuring safe and trustworthy inventory data. To monitor product movements and maximize warehouse inventories, Amazon combines blockchain technology with artificial intelligence. Blockchain data is analyzed by AI algorithms to predict changes in demand, which lowers storage costs and guarantees quicker and more effective product delivery.

#### 8.6.2.4 Case study 4: Blockchain-powered diamond tracking at De Beers

Blockchain systems with AI capabilities examine supply chain data trends to identify fraudulent supplier activity, phony invoices, and fake goods. All transactions are visible and unchangeable thanks to blockchain's immutability. To trace diamonds from mine to retail, the multinational diamond business De Beers used Tracr, a blockchain-based artificial intelligence system. AI ensures that only authentic and ethically sourced diamonds reach the market by identifying questionable transactions and synthetic diamonds. Supply chain management is being revolutionized by AI-driven blockchain monitoring, which enhances automation, fraud detection, inventory optimization, and traceability. Businesses that use blockchain solutions driven by AI lower operating costs, increase supply chain security, and boost transparency. Intelligent blockchain designs will propel the efficiency and trust of global supply chains in the future as more businesses embrace these technologies.

### 8.6.3 Healthcare (AI-powered blockchain for secure patient data)

In the healthcare sector, Blockchain technology driven by AI is transforming patient data security, medical record management, and interoperability. Problems with traditional healthcare systems, such as data breaches, fragmented patient data, and inefficient data sharing, impede patient care. These issues are resolved by blockchain technologies driven by AI, which ensure the secure, transparent, and decentralized management of medical data. The following list of case studies and important real-world examples highlights how intelligent blockchain systems are influencing the healthcare industry.

#### 8.6.3.1 Case study 1: MedRec: decentralized health records management

Blockchain technology and artificial intelligence (AI) make it possible for health data to be interoperable and impenetrable, providing accurate medical records to hospitals, insurers, and researchers while maintaining patient privacy. Blockchain ensures data integrity while AI analyzes patient data for early disease detection and customized treatments. MedRec, an MIT-backed project, handles patient medical records securely by utilizing blockchain and artificial intelligence. It gives patients authority over their health information, enabling them to provide healthcare providers selective access while maintaining data security and privacy compliance (HIPAA and GDPR).

#### 8.6.3.2 Case study 2: Pfizer and Biogen – blockchain for clinical trials

In clinical trials with blockchain, data validity and traceability are guaranteed by AI-driven blockchain technologies, which facilitate safe data sharing for medical research. This lowers fraud, improves teamwork, and speeds up personalized medicine and medication development. Blockchain and artificial intelligence are used by Pfizer and Biogen to protect clinical trial data and increase process transparency in drug development. In order to ensure data integrity and expedite medical research, AI algorithms examine blockchain-verified information to identify irregularities in trial outcomes.

#### 8.6.3.3 Case study 3: Change healthcare – blockchain-based claims processing

In the healthcare sector, fraudulent billing, insurance claims, and blockchain-based claims processing are significant obstacles. Blockchain systems with AI capabilities keep transparent and unchangeable medical billing data, which helps them identify questionable claims, validate transactions, and stop fraud. To improve the processing of medical claims, change healthcare used a blockchain network powered by artificial intelligence.

#### 8.6.3.4 Case study 4: BurstIQ – blockchain for personalized healthcare

Wearable health monitors and internet of things devices can share data securely and instantly by utilizing blockchain technology powered by artificial intelligence. AI may be used to securely store and analyze patient data, like blood sugar levels and pulse rate, to provide predictive medical insights. BurstIQ secures wearable health data by combining blockchain and artificial intelligence. Blockchain guarantees data privacy and compliance, allowing for individualized treatment recommendations, while AI algorithms examine patient health trends.

By improving patient data security, maximizing medical research, reducing fraud, and safeguarding IoT-based health monitoring, blockchain powered by artificial intelligence is transforming the healthcare industry. These intelligent blockchain designs guarantee efficiency, privacy, and trust, opening the door to a safer, data-driven healthcare system.

## 8.7 Challenges and Future Research Directions

Due to the high processing power needs of AI algorithms and the inherent limitations of blockchain networks, such as slow transaction rates and excessive resource consumption, AI-enabled blockchains have significant scaling issues. Improved interoperability, quicker consensus methods, and effective data storage are necessary for the decentralized execution of AI-driven smart contracts and machine learning. To increase performance, solutions including sharding, Layer 2 scalability, and consensus protocols tailored to AI are being investigated. Real-time AI applications on blockchain need solutions to the scalability issues to be effective, economical, and broadly embraced.

Deployment of AI on blockchain entails serious ethical and legal challenges such as data privacy, AI algorithm bias, responsibility, and compliance. Blockchain brings transparency and protection, but AI decision-making processes are typically black boxes, and fairness, prejudice, and negative impact are frequently questioned. Additionally, AI-powered smart contracts create legal ambiguity when it comes to responsibility and enforcement. Governments and regulatory agencies need to create unambiguous guidelines to regulate innovation and moral responsibility, so AI-powered blockchain systems are equitable, responsible, and in harmony with international legal norms. Blockchain 4.0 is a next-generation innovation in blockchain technology that integrates AI, automation, and cutting-edge scalability solutions for enhancing speed, efficiency, and intelligence in decentralized networks. It enables quicker transactions, self-evolving smart contracts, and evolving security protocols that position Blockchain 4.0 perfectly for implementation in finance, supply chains, Web3, and AI-driven systems. Blending self-evolving intelligence and decentralization, Blockchain 4.0 forms the basis of an autonomous, secure, and scalable digital world.

Self-adaptive AI blockchains merge blockchain technology and machine learning to develop decentralized networks that self-heal, self-optimize, and evolve. These blockchains employ AI-powered mechanisms to dynamically adjust network parameters, optimize transaction rates, and improve security through real-time detection and mitigation of threats. Through self-learning protocols, decentralized AI agents, and automated governance, they facilitate autonomous financial systems, smart supply chains, and personalized Web3 applications. By combining the adaptability of AI with the transparency and security of blockchain, self-adaptive AI blockchains open the way for a more efficient, intelligent, and decentralized digital world.

## 8.8 Conclusion

The confluence of blockchain with AI and machine learning has greatly improved scalability, security, efficiency, and interoperability within decentralized networks. With further evolution of the technologies, blockchain will become increasingly adaptive and autonomous, resulting in intelligent, self-optimizing networks. Breakthroughs in the future will revolve around AI-based consensus algorithms, quantum-resistant security, and real-time processing enhancements. Several of the breakthroughs, such as self-adaptive smart contracts, sharding, and cross-chain interoperability, already define the next-generation blockchain networks. Ethical and regulatory issues, such as AI fairness and data privacy, remain prime areas for ethical adoption. Overall, the synergy of AI and blockchain is a paradigm shift to a more secure, efficient, and intelligent decentralized digital infrastructure. Open research topics in smart blockchain architecture target increasing scalability, security, interoperability, and responsiveness through the incorporation of AI. Key directions include the creation of AI-optimized consensus algorithms such as proof-of-learning (PoL) for increased efficiency, self-tuning smart contracts for reacting to real-world data, and scalable architectures with sharding, DAGs, and AI-based load balancing. Interoperability is an open issue that demands cross-chain communication protocols and standard interfaces. Quantum-resistant cryptography, AI-based threat detection, and decentralized AI training for enhanced privacy and fraud protection are a few areas of research in security. Ethical considerations such as fairness of AI, avoidance of bias, and compliance with regulations also need to be considered. AI-based energy-efficient blockchain for green consensus patterns is another area to be researched in the future. Solving these research issues is central to the construction of secure, efficient, and intelligent decentralized systems.

## References


[1] Ben-Sasson, E., et al. (2021). "Scalability of Layer 2 Rollups in Blockchain Networks." IEEE Transactions on Blockchain.

[2] Binns, R. (2018). "Fairness in Machine Learning: Lessons for Blockchain AI Governance." Artificial Intelligence Ethics Journal.

[3] Mishra, R et al. (2022). Computational Intelligence and Blockchain-Based Security for Wireless Sensor Networks. In A. Tyagi (Ed.), *Handbook of Research on Technical, Privacy, and Security Challenges in a Modern World* (pp. 324-336). IGI Global Scientific Publishing. https://doi.org/10.4018/978-1-6684-5250-9.ch017

[4] Zhang, T., et al. (2023). "Automated Anomaly Detection in Blockchain Using AI." Cyber-Physical Systems Review.

[5] Kuznetsov, Oleksandr, et al. "On the integration of artificial intelligence and blockchain technology: a perspective about security." *IEEE Access* 12 (2024): 3881-3897.

[6] Wang, J., & Li, X. (2021). "Self-Adaptive Smart Contracts for Intelligent Blockchain Systems." Decentralized Computing Journal.

[7] Chen, L., et al. (2022). "Quantum-Resistant Cryptographic Techniques for Blockchain Security." Cryptography & Security Journal.

[8] Rizal, Syamsul, and Dong-Seong Kim. "Enhancing Blockchain Consensus Mechanisms: A Comprehensive Survey on Machine Learning Applications and Optimizations." Blockchain: Research and Applications (2025): 100302.

[9] El Hessaini, Soufiane, et al. "Enhancing Blockchain Consensus Mechanisms with Artificial Intelligence: Potential Improvements and Applications." International Conference on Digital Age & Technological Advances for Sustainable Development. Springer, Cham, 2024.

[10] Venkatesan, K., and Syarifah Bahiyah Rahayu. "Blockchain security enhancement: an approach towards hybrid consensus algorithms and machine learning techniques." Scientific Reports 14.1 (2024): 1149.

[11] Bhumichai, Dhanasak, et al. "The convergence of artificial intelligence and blockchain: The state of play and the road ahead." Information 15.5 (2024): 268.

[12] Jiang, Y., et al. (2022). "Proof-of-Learning: An AI-Optimized Consensus Mechanism for Blockchain." Journal of Distributed Ledger Technology

[13] Bello, Halima Oluwabunmi, Courage Idemudia, and Toluwalase Vanessa Iyelolu. "Integrating machine learning and blockchain: Conceptual frameworks for real-time fraud detection and prevention." World Journal of Advanced Research and Reviews 23.1 (2024): 056-068.

[14] Kuznetsov, Oleksandr, et al. "On the integration of artificial intelligence and blockchain technology: a perspective about security." IEEE Access 12 (2024): 3881-3897.

[15] Arumugam, Senthil Kumar, et al. "Quantum Computing for Next-Generation Artificial Intelligence–Based Blockchain." Quantum Computing. Auerbach Publications, 2025. 251-267.

[16] Singh, Arvind R., et al. "Optimizing demand response and load balancing in smart EV charging networks using AI integrated blockchain framework." Scientific Reports 14.1 (2024): 31768. De Filippi, P., & Wright, A. (2018). "Decentralized Blockchain Governance Models." Harvard Journal of Law & Technology.

[17] Gervais, A., et al. (2016). "On the Security and Performance of Proof-of-Work Blockchains." ACM Transactions on Privacy and Security.
[18] Hardjono, T., et al. (2020). "Interoperability in Blockchain Networks: Challenges and Solutions." MIT Media Lab Research Paper.
[19] Bera, Basudeb, et al. "AI-enabled blockchain-based access control for malicious attacks detection and mitigation in IoE." IEEE Consumer Electronics Magazine 10.5 (2020): 82-92.
[20] Udokwu, Chibuzor, Henry Anyanka, and Alex Norta. "Evaluation of approaches for designing and developing decentralized applications on blockchain." Proceedings of the 4th International Conference on Algorithms, Computing and Systems. 2020.
[21] Li, Juanjuan, et al. "Blockchain intelligence: Intelligent blockchains for web 3.0 and beyond." IEEE Transactions on Systems, Man, and Cybernetics: Systems 54.11 (2024): 6633-6642.
[22] Hussain, Imran, et al. "Proof of inherent intelligence consensus mechanism empowering blockchain-enabled transactive energy." IEEE Access (2025).
[23] Lin, Yijing, et al. "A unified blockchain-semantic framework for wireless edge intelligence enabled web 3.0." IEEE Wireless Communications (2023).
[24] X. Jiang, Y. Liu, and M. Wang, "Proof-of-Learning: A Novel AI-Driven Consensus Mechanism for Blockchain," *IEEE Transactions on Neural Networks and Learning Systems*, vol. 33, no. 5, pp. 2301–2312, May 2022.
[25] A. Singh and D. Patel, "AI-Enhanced Proof-of-Stake Protocol for Energy-Efficient Blockchains," *Journal of Blockchain Research*, vol. 8, no. 2, pp. 101–113, 2023.
[26] Y. Zhang, H. Chen, and X. Wu, "Autonomous Anomaly Detection in Blockchain Using Deep Reinforcement Learning," *IEEE Access*, vol. 11, pp. 17642–17652, 2023.
[27] T. Wang and J. Li, "Self-Adaptive Smart Contracts Using AI and Formal Verification," *Future Generation Computer Systems*, vol. 115, pp. 52–64, 2021.
[28] V. Buterin, "Ethereum 2.0 and Sharding: Technical Overview," *Ethereum Foundation Blog*, 2020.
[29] S. Popov, "The Tangle: A DAG-Based Ledger for Internet of Things," *IOTA White Paper*, 2017.
[30] E. Ben-Sasson et al., "Scalable, Transparent, and Post-Quantum Secure Layer 2 Rollups," *zkSync White Paper*, 2021.

[31] T. Hardjono, N. Smith, and A. Pentland, "Towards a Design for the Blockchain Interoperability Architecture," *MIT Connection Science Technical Report*, 2020.

[32] L. Zhao, J. Wang, and F. Liu, "Cross-Chain Communication Protocols for Blockchain Networks," *IEEE Communications Surveys & Tutorials*, vol. 23, no. 1, pp. 106–128, 2021.

[33] R. Binns, "Fairness in Machine Learning: Lessons from Political Philosophy," in *Proceedings of the 2018 Conference on Fairness, Accountability and Transparency (FAT)*, pp. 149–159, 2018.

[34] G. Zyskind, O. Nathan, and A. Pentland, "Decentralizing Privacy: Using Blockchain to Protect Personal Data," in *2015 IEEE Security and Privacy Workshops (SPW)*, pp. 180–184.

[35] B. Mittelstadt, P. Allo, M. Taddeo, S. Wachter, and L. Floridi, "The Ethics of Algorithms: Mapping the Debate," *Big Data & Society*, vol. 3, no. 2, pp. 1–21, 2016.

[36] A. Savelyev, "Contract Law 2.0: 'Smart' Contracts as the Beginning of the End of Classic Contract Law," *Information & Communications Technology Law*, vol. 26, no. 2, pp. 116–134, 2018.

[37] P. De Filippi and A. Wright, *Blockchain and the Law: The Rule of Code*, Harvard University Press, 2018.

[38] M. Schröder, "Blockchain and Securities Regulation in the EU," *Capital Markets Law Journal*, vol. 14, no. 3, pp. 329–344, 2019.

[39] C. Harvey, "Taxation and the Blockchain: Key Issues," *Journal of Taxation and Regulation of Financial Institutions*, vol. 33, no. 4, pp. 18–25, 2020.

[40] C. Catalini and J. Gans, "Initial Coin Offerings and the Value of Blockchain Transparency," *NBER Working Paper No. 24418*, 2019.

[41] F. Wang, J. Chen, and S. Wang, "Regulatory Sandboxes for Blockchain: Opportunities and Challenges," *Technology in Society*, vol. 62, pp. 101–115, 2020.

[42] I. Pernice, K. Gatteschi, and S. Lamberti, "Blockchain Regulation and Governance in Europe: Co-Creation of Best Practices," *European Journal of Law and Technology*, vol. 11, no. 3, 2020.

[43] L. Chen, S. Jordan, Y.-K. Liu, D. Moody, R. Peralta, R. Perlner, and D. Smith-Tone, "Report on Post-Quantum Cryptography," *NIST IR 8105*, National Institute of Standards and Technology, 2022.

[44] Q. Yang, Y. Liu, T. Chen, and Y. Tong, "Federated Machine Learning: Concept and Applications," *ACM Transactions on Intelligent Systems and Technology*, vol. 10, no. 2, pp. 1–19, 2019.

[45] Y. Kou, Y. Peng, Y. Chen, and Z. Xie, "AI-Powered Compliance Monitoring for Financial Blockchain Applications," Journal of Financial Technology, vol. 3, no. 1, pp. 33–47, 2022.

[46] R. Sharma, N. Gupta, and H. Singh, "Automated Cybersecurity Frameworks Using AI for Blockchain Systems," IEEE Transactions on Industrial Informatics, vol. 19, no. 2, pp. 1655–1664, 2023.

[47] Li, Juanjuan, et al.(2024) "Blockchain intelligence: Intelligent blockchains for web 3.0 and beyond." *IEEE Transactions on Systems, Man, and Cybernetics: Systems* 54.11 (2024): 6633-6642.

[48] Mishra, R., Narwaria, M., Singh, M., Dwivedi, V., Kaur, I., Saxena, S. (2026). Blockchain-Based Authentication in 6G Metaverse Using Hyperledger Indy for Self-Sovereign Identity (SSI). In: Kumar, A., Rathore, P.S., Moreira, F. (eds) Blockchain Enabled Metaverse for Smart Wireless Sensor Networks. Information Systems Engineering and Management, vol 72. Springer, Cham. https://doi.org/10.1007/978-3-032-07305-1_10

[49] Mishra, R., Kaur, I., Sharma, V., & Bharti, A. (2022). Computational Intelligence and Blockchain-Based Security for Wireless Sensor Networks. In A. Tyagi (Ed.), Handbook of Research on Technical, Privacy, and Security Challenges in a Modern World (pp. 324-336). IGI Global Scientific Publishing. https://doi.org/10.4018/978-1-6684-5250-9.ch017

[50] Mahesh, G., & Renu, M. (2025). PCBQC: A Blockchain-Based, Patient-Centric EHR Management Framework using Hybrid Post-Quantum Lattice Cryptographic Algorithms. Int. J. Perform. Eng., 21, 617.

# 9

# Shifting of Origin and Error Minimization for Realistic Representation of Basic Geometrical Entities on Computer Screen

**S. C. Dimri[1] and R. Indu[2]**

[1,2]Department of Computer Science and Engineering, Graphic Era Deemed to be University, India

**Email:** dimri.sushil2@gmail.com; richaindu@outlook.com

## Abstract

In today's fast-paced digital era, there is a remarkable shift as digital technology and screens increasingly take the place of traditional books and printed materials. A computer screen is a raster of pixels, whereas a chapter is a collection of an infinite number of points; thus, to map a realistic geometric object on the computer screen can be seen as a mapping from an infinite set to a set of a few pixels. And, due to this, the real representation of geometrical objects on a computer screen is almost impossible, but with some basic methods, the deformation and error in the representation of geometrical shapes on the screen can be minimized. This chapter discusses some basic mathematics, which is used in computer graphics to minimize errors in the representation of geometrical shapes. These methods are effective and need further enhancements to achieve good results.

## 9.1 Introduction

A discipline of computer science that focuses on the mathematical interpretations of algorithms by describing them in terms of geometry is referred to as computational geometry. Generally, computational geometric approaches are susceptible to a couple of purely geometrical concerns, which are also



DOI: 10.1201/9788743812739-9

considered as a part of computational geometry. Although computational geometry is a recent field, it is one of the oldest in computer science, with origins extending back to antiquity [1]. Computational complexity is at the heart of computational geometry, and it has substantial practical ramifications when methods are employed for massive datasets involving tens or hundreds of millions of points. The time difference between $O(n_2)$ and $O(n \log n)$ with these collections may be quantified in days rather than seconds [2].

For the description and change of three-dimensional geometric entities in computer graphics, a representational model is put forward. Dual unit vectors constitute the basis of the suggested representation's structure, and dual unit quaternions or dual orthogonal matrices are used to perform the associated transformations. Since the other relevant geometric properties of a represented curve or surface, such as a tangent or normal vector, are integrated inside the actual representational structure itself, the fundamental advantage of this representation is its compactness. The notion of screw displacement is a logical way to define rotations, translations, and perspective transformations; the moment vector of each dual line is used to achieve scaling.

The study of an object's size, shape, and location in space is known as object geometry. Mathematical models like points, lines, and planes can be used for this [3]. In order to comprehend pixel addressing and object geometry in computer graphics, a solid understanding of raster is first required. A raster is a rectangular array of cells, each with its own unique address [4]. Two-dimensional pictures like bitmaps are represented using this addressing system. The two most popular techniques for addressing pixels in a raster are linear addressing and coordinate indexing. Simply giving each pixel a series of addresses, starting at 0, is linear addressing. Each pixel is given a $(x,y)$ coordinate pair that describes its location within the raster as part of the coordinate indexing process. The next step after taking care of the pixels is to decide how to use them to draw objects on the screen [5].

### 9.1.1 Applications

The ideas of object geometry and pixel addressing have a wide range of possible applications in computer graphics. These ideas, for instance, can be used for the development of novel techniques for altering and interactively viewing digital photographs, as well as for the improvement of image processing algorithms. Additionally, computer graphics designers may develop more logical and user-friendly interfaces for their goods by comprehending how pixel addressing and object geometry interact [6].

### 9.1.2 Advantages

Pixel addressing and object geometry in computer graphics have the benefit of enabling the creation of more realistic visuals. Computer graphics designers may produce graphics that are more faithful representations of reality by comprehending how pixels are addressed and how objects are positioned with respect to one another. Pixel addressing and object geometry also enhance the efficiency of computer graphics programs. Computer graphic designers may develop algorithms that more effectively use the system's resources, leading to quicker rendering speeds and smoother animations, by knowing how pixels are addressed and how objects are positioned in relation to one another. Finally, object shape and pixel addressing can also aid in the development of user-friendly interfaces. By being aware of how pixels work and how objects are positioned with each other, computer graphic designers can create interfaces that are easier to use and navigate.

### 9.1.3 Disadvantages

Computer graphics involves pixels for portrayal. Pixels are important for a lattice that is planned for the screen, which can show pictures on the screen. This planning is called pixel addressing and is generally utilized because of its minimal expense and excellent goal. Pixel tending, too, has a couple of inconveniences. One of them is that it offers no natural mathematical properties at the pixel level, which is significant in certain applications. For instance, while registering with pixels, one can only with significant effort decide the slant or region of an item in a picture without scanning across numerous pixels and knowing the size of every pixel.

## 9.2 Related Work

Guru and Nagendraswamy [7] suggested an approach of shape encoding and retrieval, which was demonstrated to be resilient to small deformations and occlusions as well as invariant to picture transformations like scaling, translation, reflection, and rotation. Numerous trials have also been carried out to show the methodology's viability and to highlight its advantages over other methodologies. Westheimer [8] questioned the possibility that the transition between the two spaces can be usefully examined in terms of differential geometry and metrical qualities due to the obvious separation between object and visual space that they indicate. Kynigos and Latsi [9] examined that the software's operations and meanings are intertwined with abilities like spatial orientation and spatial visualization. Although the body-syntonic metaphor continues to be important for steering the turtle in the three-dimensional (or 3D-) virtual environment, it appears that it must be coordinated with other frames of reference that frequently clash with one another. Additionally, the strong relationship between spatial graphical and geometrical components highlighted by the software's dragging functionalities allowed students to progress beyond an initial perceptual approach. Enabling them to link geometrical figures to real 3D objects, as well as the shifting of planes in 3D space to realistic angle situations. This framework employs the concept of angle as a turn and measure with an emphasis on directionality, as well as focusing on a connection among 2D representations of different planes.

Zhu and Wu [10] investigated in-plane geometric deviations, where common AM deviation causes may be related to various transformations, such as translation, rotation, and scaling of the nominal shape. They revealed that by defining transformation parameters, analytical deviation models of regular shapes can be generated in the deviation space, which are also capable of being calibrated using measurement data. Furthermore, to demonstrate the viability of the suggested strategy, a case study was also conducted. Nordberg [11] proposed that a two-dimensional plane in Euclidean geometry is made up of points, lines, circles, ellipses, and a variety of curves, including parabolas, hyperbolas, and spirals. The attributes associated with these geometric objects include angles, areas, and distances (or lengths). Euclidean geometry examines how individual objects relate to one another and how they interact. For instance, intersections of lines or curves and the incidence connections between points, sets of points, and lines, or more general curves. Euclidean geometry is also applicable to objects in 3D space, comprising curves, surfaces, solids, planes, as well as points and lines. These objects' attributes in 3D space also include volumes and solid angles. One way to think of trigonometry is as a particular branch of geometry that focuses on angles. Spröte et al. [12] find that an object's causal origins, that is, the processes that shaped it in the past, have a major impact on shape representations as well. To describe the apparent symmetry axis of the items, observers put dots on them. Responses closely resembled the object's geometrical axes when it seemed "complete" – made wholly by a single generating process.

Wang [13] produced the aforementioned questions, which yielded positive results from his investigation. A unique geometrical holographic display that reproduces realistic 3D pictures with all of the human visual cues and without any visual side effects has been presented. It leverages geometrical optical principles. A least essential light field representation was further created, which can offer suggestions on how to reduce the amount of data required when developing an FP3D or creating 3D scenes. A proof-of-concept prototype that can deliver realistic 3D sceneries with depth ranges greater than 5m is then put up. Bonnen et al. [14] show that the tuning of cortical neurons and relevant features of perception are fundamentally shaped by this mapping. The concept explains startlingly non-canonical tuning in the available electrophysiological evidence for 3D motion, as well as different patterns of perceptual mistakes seen in human behavior. The geometry connecting the environment to the sensory epithelium has a significant impact on decoding the world from cerebral activity. Similarly, Đuri´ et al. [15] proposed a method for producing 2D CAD-like visualizations of complicated geometry using photogrammetric models. This extends aid to identifying contours of specific object elements during 2D shape analysis. Further, this technique leverages point clouds obtained from photogrammetric models to build a planar representation of the object's geometry by segmenting points depending on the vertical geometric characteristics.

Jiang et al. [16] proposed two types of anti-aliasing approaches for computer graphics: pre-filtering-based anti-aliasing and post-filtering-based anti-aliasing. Both techniques are thoroughly reviewed in their work. This divides post-filtering-based anti-aliasing algorithms into two further categories: post-process methods for delayed rendering and hardware anti-aliasing solutions. Sills and Williams [17] suggested a novel method that takes advantage of graphics hardware functions, which is discussed and contrasted with preceding methods. It is demonstrated that the method reduces the amount of aliasing just as well while being quicker and less difficult. Zhao et al. [18] showed the SMORE algorithm's potential for usage in both research and clinical settings. This paper first discusses it before demonstrating its performance in four applications. It is initially demonstrated to enhance the visibility of brain white matter lesions in FLAIR images obtained from individuals with multiple sclerosis. It is then demonstrated to enhance scarring visualization in cardiac left ventricular remodeling following myocardial infarction. An Abroad framework for multiresolution neural networks and a structure of imaging application, named MR-Net, was suggested by Paz et al. [19]. Their coordinate-based networks consist of numerous phases that incrementally incorporate more fine-grained information, making them continuous in both space and scale. They are also a neat and effective depiction. This paper provides programs for texture magnification, minification, and antialiasing as well as instances of multiresolution picture representation.

Rajarapollu and Mankar [20] examined the effectiveness of morphological anti-aliasing, super-sampling, and multi-sample approaches. Although super-sampling and multi-sampling have been found to produce superior outcomes, they are also more computationally intensive. Morphological anti-aliasing techniques have the potential to enhance performance while still producing real-time results. Originally, Dimri and Ram [21] worked toward reducing the complexity of the Midpoint ellipse algorithm. To this purpose, they estimated the pixels of the ellipse in an octant and plotted the remainder of the curve with the reflection around the line. Then, to visualize every pixel on the ellipse, the parametric equation of the auxiliary circle was implemented. In another work, Dimri et al. [22] suggested a novel recursive method to perform scan conversion of the circle, which computed each pixel in an octant while the remainder of the circle was constructed using symmetry. Kumar et al. [23] and Saini [24] did some more work in 2D and 3D representation of geometrical objects on a computer screen with more accuracy and the least aliasing effects.

## 9.3 Realistic Representation and Error Minimization

There is a difference between a plain paper sheet and a computer screen. A computer screen is a matrix of pixels, while plain paper is a collection of infinite points. Pixel coordinates are always integer-valued. Thus, drawing a geometrical shape on a computer screen may contain a lot of errors, because this collection of infinite points was to be mapped to a few pixels on a computer screen. Generally, a Pixel is an existing entity (picture cell) that has height and width, whereas a point has theoretical existence. Not only this, but computer coordinate systems and Cartesian coordinate systems are also different. In a computer coordinate system, this paper has only one quadrant to draw a geometrical shape, while Cartesian coordinates exist in more than one positive quadrant. It is not possible to draw on a computer screen.

If this paper draws basic shapes like a line, circle, ellipse, or parabola that exist in more than one quadrant, this paper needs to shift the origin of the computer coordinate system somewhere in the middle of the screen. Thus, shifting of origin and error minimization is a must to draw either simple or complex geometrical shapes on the computer screen. Error minimization in drawing the figure while drawing it on screen is important to represent the figure more realistically on the computer screen. In the Digital Differential Analyzer (DDA) algorithm, the error minimization is achieved with the help of a round function, which rounds the real point to the nearest integer pixel position.

Using the DDA line algorithm, for slope ($0 < m < 1$),

$$Xn + 1 = Xn + 1 \tag{9.1}$$

$$Yn + 1 = Yn + m. \quad \text{(non-integer value)} \tag{9.2}$$

Where, Round (4.79) = 5, and Point (5, 4.79) is equivalent to (5, 5) pixels. Now, increasing the resolution of the screen is a simple solution to the problem of getting a smooth drawing of geometrical shapes, but it is not that effective.

In Bresenham's line, in each step of pixel determination, select that pixel, that is, much closer to the real theoretical point, and this may achieve error minimization. Similarly, in circle generation, that pixel is selected as the next pixel that is much closer to the circumference of the circle. For slope, $0 < m < 1$, Endpoints of the line segment, which want to draw on the computer screen $A(x_1, y_1)$ and $B(x_2, y_2)$, slope $m$ can be evaluated as:

$$M = (y_2 - y_1)/(x_2 - x_1) \tag{9.3}$$

or,

$$m = \Delta y/\Delta x. \tag{9.4}$$

Likewise, in the Bezier curve drawing, the shape of the curve is based on the control point, and the curve has many non-integer points, which are then changed to the nearest pixels by using the round function.

### 9.3.1 Shifting of origin

It is necessary to draw a figure in full shape on the computer screen, which has only one quadrant. In Figure 9.1, coordinates for P′ are $(h+x, k-y)$ w.r.t the computer coordinate system. So, to plot the P′, the command must be put pixel $(h+x, k-y, 3)$; this will plot pixel P′ with coordinate $(x, y)$ in a new coordinate system, which has its origin at point $(h, k)$ concerning the computer coordinate system. This shifting provides the facility of all four quadrants, and thus any geometrical shape can be drawn. Pixel $P(x, y)$ and $P'(x, y)$ have some coordinates and nature, but one has the coordinates with the computer coordinate system, whereas the second one has the same with respect to the new coordinate system. So, $P'(x, y)$ is equivalent to the similar pixel at pixel $P(x, y)$.

Similarly, in circle drawing, whether using Bresenham's or mid-point algorithm (as shown in Figure 9.2), find pixels only in octant $45° \leq \theta \leq 90°$, and the rest parts of the circle will be generated by reflection about the $x$-axis, $y$-axis, line $y = x$, and line $y = -x$.

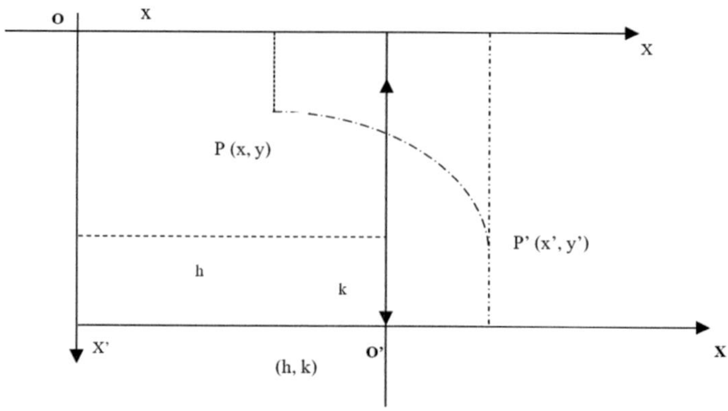


**Figure 9.1** Shifting of origin to create four quadrants on the computer screen.

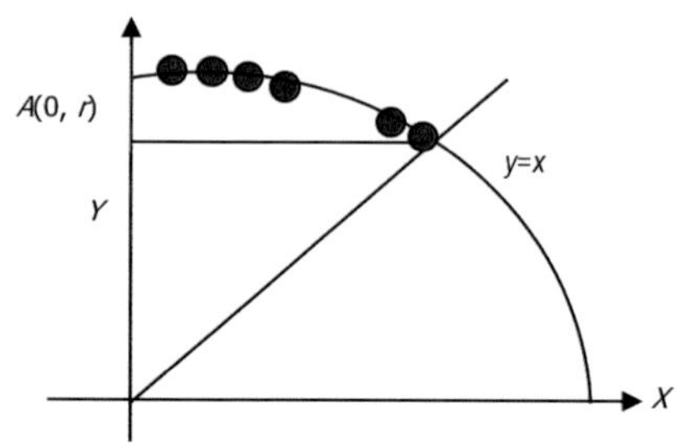


**Figure 9.2** Pixel determination along the arc of circle algorithm given by Dimri et al. [22].

From Figure 9.3, *A* (0, *r*) is the starting point, from *A* to *L* along the arc of the circle, and need to determine the pixels. While moving *A* to *L* along the arc of the circle, the change in *x*- and *y*-coordinates is:

$$\Delta x = \left(\frac{r}{\sqrt{2}} - 0\right) = \frac{r}{\sqrt{2}} = 0.7072r \tag{9.5}$$

and,

$$\Delta y = \left(r - \frac{r}{\sqrt{2}}\right) = r\left(1 - \frac{1}{\sqrt{2}}\right) = 0.2928r. \tag{9.6}$$

Clearly $\Delta x > \Delta y$, the change in the *x*-coordinate is greater than the *y*-coordinate. This maximum possible increment of 1 will be given to the *x*-coordinate, and the value of the *y*-coordinate will be decided by the decision parameter. To minimize the error in drawing a circle on the screen, select the pixel between

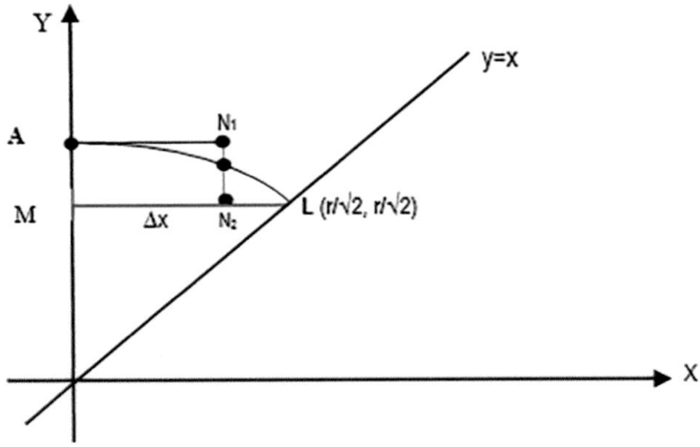


**Figure 9.3** In a circle drawing, determination of the next pixel close to the real theoretical point *R*.

$N_1$ and $N_2$, which is close to the theoretical circumference of the circle. Thus, deciding which coordinate component will be incremented by 1.

Ellipse generation faces the same type of problem in determining the next pixel. Again, the round function selects that pixel, which is close to the real theoretical point, as demonstrated in Figure 9.5.

While generating an ellipse with the help of an auxiliary circle, the pixels are determined in the octant $45° \leq \theta \leq 90°$ and the pixels in the octant $0° \leq \theta \leq 45°$ are determined by taking the reflection of the pixels in the octant $45° \leq \theta \leq 90°$ about the line $y = x$. As we know that the reflection of a point $(a, b)$ about the line $y = x$ is given by $(b, a)$, as shown in Figure 9.4.

Then, by using the equations $x^2 + y^2 = a^2$ (auxiliary circle) and the parametric equation of circle: $x = a\cos\phi$ and $y = a\sin\phi$

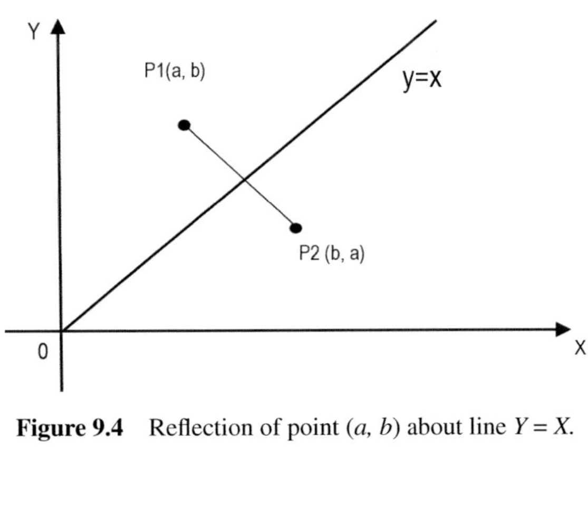


**Figure 9.4** Reflection of point $(a, b)$ about line $Y = X$.

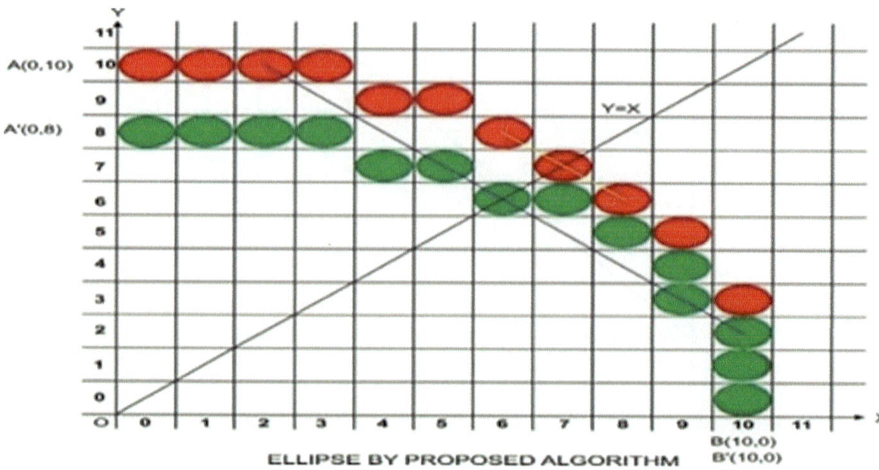


**Figure 9.5** Ellipse generation by the algorithm given by Dimri and Ram [21] and plotting of pixels on the Cartesian grid.

To find corresponding points on an ellipse using the conversion

$$x' = x = a\cos\phi \text{ and } y' = \frac{b}{a}(y) = \frac{b}{a}\ (a\sin\phi),$$

$$x' = a\cos\phi \text{ and } y' = b\sin\phi, \text{ the parametricequationof ellipse}$$

where $(x', y')$ is the point on the ellipse corresponding to point $(x, y)$ on the circle, now using the above equations (Figure 9.5).

This chapter has the pixels for the entire quadrant of the ellipse, since the ellipse is symmetrical about the $x$-axis and the $y$-axis, so by taking a reflection about the $x$-axis and the $y$-axis, we will get the complete ellipse. This method is far simpler than the midpoint ellipse generation algorithm and is computationally very low [22].

### 9.3.2 Role of midpoint and $f(x, y) = x^2 + y^2 - r^2$ in circle generation

In Bresenham's and midpoint, circle generation algorithm, we use function $f(x, y) = x^2 + y^2 - r^2$ to locate the position of a point with respect to a circle, on the basis of the sign of $f(x, y)$, the algorithm decides which pixel is close to the real. Theoretical point $R$, in midpoint circle, the position of midpoint $M$ (midpoint of $N_1N_2$) helps to determine which pixel is close to the circle, and this way the algorithm minimizes the error as shown in Figure 9.6.

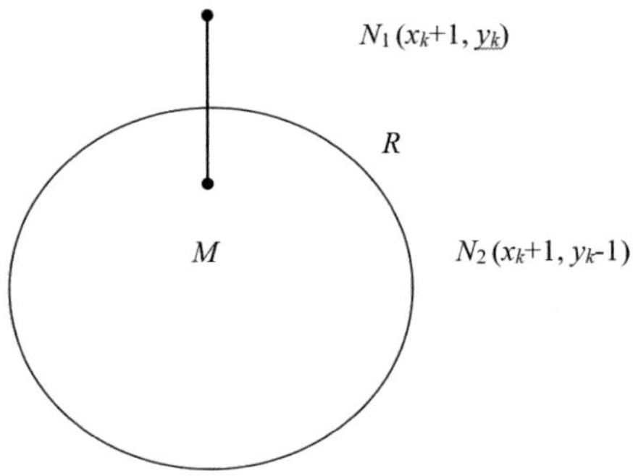


**Figure 9.6** Role of midpoint to decide next pixel position in midpoint algorithm.

If $M$ is an interior point of a circle, then $N_1$ is close to Circle, otherwise $N_2$,

$$N_1M = N_2M$$

$$N_1R = \left\{\left(\frac{N_1N_2}{2}\right) - \delta\right\}$$

$$N_2R = \left\{\left(\frac{N_1N_2}{2}\right) + \delta\right\}$$

where, $RM = \delta$ (says) ($\delta > 0$) since $N_1R < N_2R$, so $N_1$ is close to the circle circumference.

### 9.3.3 Bezier curve/surface generation

The Bezier curve is given by eqn (9.7) and represented in Figure 9.7.

$$p(u) = \sum_{k=0}^{n} p_k C_k^n (u)^k (1-u)^{n-k} \tag{9.7}$$

where $0 \le u \le 1$, and the curve's nature is controlled by control points. The Bezier curve passes through the first and last control points; the rest in between have infinite points corresponding to the value of $u$, such that $0 \le u \le 1$. To find the pixel position representing the curve on the screen, the round function is used to minimize the error for non-integer points shown in eqn (9.8) and (9.9).

$$x(u) = \sum_{k=0}^{n} x_k C_k^n (u)^k (1-u)^{n-k} \tag{9.8}$$

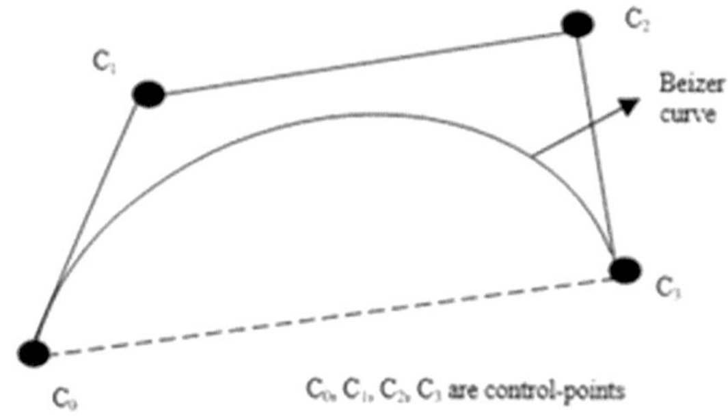


**Figure 9.7** Generation of a non-smooth cubic Bezier curve, where $C_0$, $C_1$, $C_2$, and $C_3$ are control-points.

$$y(u) = \sum_{k=0}^{n} y_k C_k^n (u)^k (1-u)^{n-k} \tag{9.9}$$

Yet again, $0 \leq u \leq 1$. For infinite values of *u,* there are a finite number of points on the computer screen which distort the shape of the curve. If control points are in different quadrants, then for curve generation, we need to shift the origin to a new position so as to complete the formation of all four quadrants. Again, for surface generation, we need orthogonal Bezier curves, such that both, $0 \leq u \leq 1$ and $0 \leq v \leq 1$:

In cubic Bezier surface patch generation, the approximation of the point is toward the nearest pixel positions. To minimize the error, the non-integer points are again rounded to the nearest integer. This brings a lot of deformation in the shapes of geometrical objects when drawn on the screen. To overcome this problem, a simple solution is to increase the resolution of the screen. But for ideal situations, the resolution of the screen must be infinite, which is not possible; thus, fine and smooth figure generation on the screen is an almost impossible task. But by using error minimization and increasing the resolution, this problem can be brought under control.

Matrix of control points: which controls the shapes of orthogonal Bezier curves that form the cubic Bezier surface patches shown in Figure 9.8.

To plot the pixels corresponding to points on the Bezier curves, the coordinate values need to be rounded to the nearest integers, which creates a Zigzag edge pattern and results in a rough, blurred surface patch due to aliasing, which creates a problem when these surface patches combine and form a surface terrain.

An anti-aliasing method can be used to minimize the effect of the zigzag, but these are not capable removing this staircase problem completely.

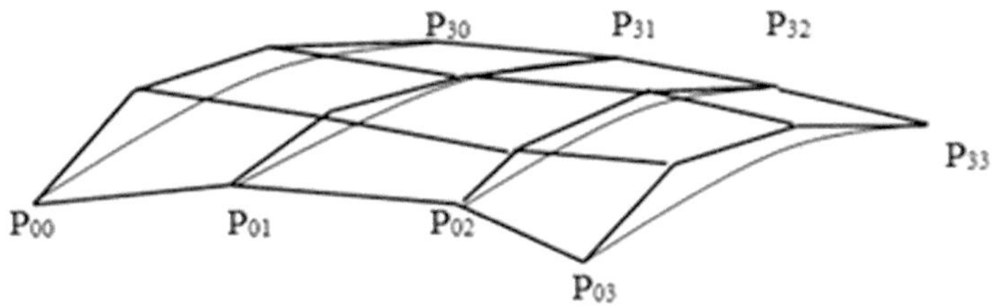


**Figure 9.8** Generation of a non-smooth cubic Bezier surface patch.

## 9.4 Conclusion

Several methods in computer graphics exist to minimize errors and aliasing, etc., but they are not very effective, since all these work on edges and color intensities of the pixels. This chapter tells us about the minimization of error using basic mathematics. More sophisticatedly, the algorithm can reduce the complexity and errors in the realistic representation of geometrical objects on the computer screen. This chapter suggests the selection of those pixels that are close to real theoretical points. Further, picture quality improves with an increase in resolution. Thus, it has more pixels where points can be mapped, and error dimensions can be reduced significantly. In this way, with sufficiently good resolution and error minimization methods, one can draw the geometrical shapes on the computer screen in a more realistic way. The mathematics and high-resolution screen can give us better control over the quality of the image, which is what is generated on the computer screen. A tricky calculation in next pixel determination and efficient recursive equation design gives us the best fast results. It is impossible to have a screen with infinite resolution; thus, the best strategies for error minimization are very important. The chapter discusses how to use shifting of origin and error minimization to generate a complete figure of a geometrical entity on the computer screen. The mapping of points and the shifting origin gives us the opportunity to draw the complete figure on the computer screen.

## References

[1] Mount, D.M. CMSC 754 Computational Geometry. Department of Computer Science, University of Maryland Fall, 2002. Retrieved from https://www.cs.umd.edu/~mount/754/Lects/754lects.pdf.

[2] Sarker, I. H. Deep Learning: A Comprehensive Overview on Techniques, Taxonomy, Applications and Research Directions. SN Computer Science, 2(6), 2021.

[3] Taffesse, W., & Kassa, L. Engineering Drawing. Haramaya University in collaboration with the Ethiopia Public Health Training Initiative, The Carter Center, the Ethiopia Ministry of Health, and the Ethiopia Ministry of Education, 2005. Retrieved from https://www.cartercenter.org/resources/pdfs/health/ephti/library/lecture_notes/env_health_science_students/engineeringdrawing.pdf

[4] Eck, D.J. Introduction to Computer Graphics. Hobart and William Smith Colleges, Version 1.4, 2023.

[5] Grande, J. C. Principles of Image Analysis. Metallography, Microstructure, and Analysis, 1(5),227–243, 2012.

[6] Krestanova, A., Cerny, M., & Augustynek, M. Review: Development and Technical Design of Tangible User Interfaces in Wide-Field Areas of Application. Sensors, 21(13), 2021.

[7] Guru, D. S., & Nagendraswamy, H. S. Symbolic representation of two-dimensional shapes. Pattern Recognition Letters, 28(1), 144–155, 2007.

[8] Westheimer, G. Illusions in the spatial sense of the eye: Geometrical–optical illusions and the neural representation of space. Vision Research, 48(20), 2128–2142, 2008.

[9] Kynigos, C., & Latsi, M. Turtle's Navigation and Manipulation of Geometrical Figures Constructed by Variable Processes in a 3d Simulated Space. Informatics in Education, 6(2), 359-372, 2007.

[10] Zhu, J., & Wu, P. A Common Approach to Geo-Referencing Building Models in Industry Foundation Classes for BIM/GIS Integration. ISPRS International Journal of Geo-Information, 10(6), 2021.

[11] Nordberg, K. Introduction to Representations and Estimation in Geometry. Computer Vision Laboratory, Department of Electrical Engineering Linkoping University, Version: 0.40, 2018.

[12] Spröte, P., Schmidt, F., & Fleming, R. W. Visual perception of shape altered by inferred causal history. Scientific Reports, 6(1), 2016.

[13] Wang, G. Geometrical holographic display, 2021. Retrieved from https://arxiv.org/ftp/arxiv/papers/2111/2111.01909.pdf

[14] Bonnen, K., Czuba, T. B., Whritner, J. A., Kohn, A., Huk, A. C., & Cormack, L. K. Binocular viewing geometry shapes the neural representation of the dynamic three-dimensional environment.Nature Neuroscience, 23(1), 113–121, 2019.

[15] Đurić, I., Obradović, R., Vasiljević, I., Ralević, N., & Stojaković, V. Two-Dimensional Shape Analysis of Complex Geometry Based on Photogrammetric Models of Iconostases. Applied Sciences, 11(15), 2021.

[16] Jiang, X., Sheng, B., Lin, W., Lu, W., & Ma, L. Image anti-aliasing techniques for Internet visual media processing: a review. Journal of Zhejiang University SCIENCE C, 15(9),717–728, 2014.

[17] Sills, T. G., & Williams, O. M. Aliasing and scintillation reduction in real-time computer graphics.; In R. L. Murrer, Jr. (Ed.), SPIE Proceedings-Optical Engineering, 43(8), 247-25, 2003.

[18] Zhao, C., Shao, M., Carass, A., Li, H., Dewey, B. E., Ellingsen, L. M., Woo, J., Guttman, M. A., Blitz, A. M., Stone, M., Calabresi, P. A., Halperin, H., & Prince, J. L. Applications of a deep learning method for anti-aliasing and super-resolution in MRI. In Magnetic Resonance Imaging (Vol. 64, pp. 132–141, 2019.

[19] Paz, H., Novello, T., Silva, V., Schardong, G., Schirmer, L., Chagas, F., Lopes, H., & Velho, L. Multiresolution Neural Networks for Imaging. In 35th SIBGRAPI Conference on Graphics, Patterns and Images, pp. 174-179, 2022. https://doi.org/10.1109/sibgrapi55357.2022.9991765

[20] Rajarapollu, P. R., & Mankar, V. R. Design and Analysis of Anti-Aliasing Filters for Rendering of Graphical Entities. In Proceedings of the Second International Conference on Information and Communication Technology for Competitive Strategies (ICTCS), 70, 1-5, 2016. https://doi.org/10.1145/2905055.2905277

[21] Dimri, S.C., & Ram, M. An efficient algorithm to scan conversion of ellipse under auxiliary circle. Nonlinear Studies, 24(1), 2017.

[22] Dimri, S.C., Chamoli, S.K., & Dimri, A. A New Dynamic Programming Approach for Scan Conversion of a Circle. International Journal of Computer Applications, 50(17), 7-10, 2012.

[23] Kumar, B., Tiwari, U. K., Kumar, S., Negi, H. S., & Kumain, K. Two dimensional successive transformations in computer graphics: Mathematical and matrix notation. In International Scientific and Practical Conference "Technology In Agriculture, Energy and Ecology" (TAEE2022). AIP Publishing, 2022. https://doi.org/10.1063/5.0103814

[24] Saini, D. Inverse Reconstruction Methodologies: A Review, Applications and Applied Mathematics, 17(3), 2022.

# 10

# Big Data Analytics using Computational Intelligence

**Amritansh Singh[1], Monali Gulhane[2], Rashmi Sharma[3], Nitin Rakesh[4] and Gunjan Chhabra[5]**

[1,2,4] Symbiosis Institute of Technology, Nagpur Campus, Symbiosis International (Deemed University), India
[3]SVKMs NMIMS University, Mukesh Patel School of Technology Management and Engineering, Shirpur campus, Dhule, Maharashtra, India
[5]School of Science and Technology, Swami Rama Himalayan University, India

**Email:** amritanshsingh2085@gmail.com; monali.gulhane4@gmail.com; rashmi.nov30@gmail.com; nitin.rakesh@gmail.com; chhgunjan@gmail.com

## Abstract

With increasing volumes of data and computational power, information must be analyzed through big data analytics so that sectors can better protect themselves against cybersecurity threats and manage risk. This chapter will discuss various computational techniques for intelligence, including artificial neural networks, quantum neural networks, and generative adversarial networks, in strengthening cybersecurity frameworks. By computational intelligence, we are able to talk about how we tackle the problems that are very complex in the cyber world and resolve them through behavioral analysis and anomalies within large-scale systems and critical infrastructure, such as aviation networks. This chapter depicts how computational intelligence is brought into the scheme of integration with cybersecurity while at the same time establishing the ability of computational intelligence toward adaptability and scalability in real-time threat mitigation. The ability of organizations through Computational Intelligence to make their cybersecurity systems both



DOI: 10.1201/9788743812739-10

effective and flexible makes it possible for proactive management of evolving threats in the digital world.

## 10.1 Overview of Big Data

### 10.1.1 Introduction to big data

Big data refers to the datasets that are large, fast, and complex to be processed using traditional data processing techniques. It is created from volumes of data generated by different sources such as social media, financial transactions, sensors, and IoT devices. Sometimes, it is credited with the five Vs: Volume, Velocity, Variety, Veracity, and Value [1], [2]. It indeed has changed the way organizations process, analyze, and apply information to make decisions. The increasing volume of digital data poses great challenges but also loads of opportunities for industries to look into the behavior of customers, optimize their processes, and enhance decision-making with advanced analytics [3–5].

A key element in big data is changing the business and institutional models in their work, and it can be applied in innovations in healthcare, finance, transportation, and cybersecurity. However, the tremendous amount of data makes an enormous challenge in terms of capacity in terms of storage, processing, and analysis, giving rise to specific tools and technologies designed to cope with such complexities.

### 10.1.2 The big data environment

The big data environment is very broad and encompasses tools, technologies, and infrastructure that support the handling of huge volumes of data. At its core, the Big Data environment is anchored on distributed computing frameworks [6], [7] which enable data to be processed across multiple machines or clusters, hence enabling scalable handling of large-scale data. Following are the main pillars of of Big Data Environment that represents in Figure 10..1:

- Cloud computing: The cloud offers scalable storage and computational resources [8], [9]; it so happens that this is one prerogative for the management of Big Data. By way of these cloud platforms – Amazon Web Services (AWS), Google Cloud, for example, or Microsoft Azure – organizations can store and process substantial data sizes without a huge investment in on-premises high-end infrastructure.
- Hadoop ecosystem: Apache Hadoop is a distributed framework that enables the storage and processing of big data across clusters of computers [10], [11]. HDFS offers scalable data storage, and MapReduce allows for the parallel processing of data.

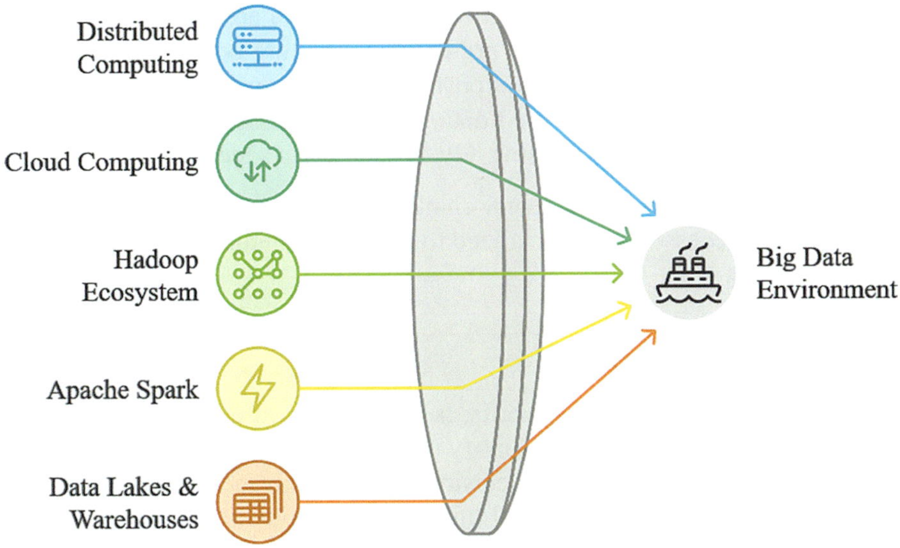


**Figure 10.1** Big data environment.

- Apache spark: This is one of the most widely used distributed computing frameworks, which aims to speed up data processing with in-memory computation. It is often much preferred over Hadoop whenever real-time data analytics is involved, owing to its excellent processing rate and ease of use [12], [13].

Big data environments embrace data lakes and data warehouses that support the storage of structured and unstructured data, allowing them to be useful in easier handling and querying large amounts of information.

### 10.1.3 Types of big data

Big data can be categorized into three primary types:

- **Structured data:** This is one of the most structured forms of data: it is searchable, as well as well-arranged [3], [14]. They are usually found in relational databases storing data with a predefined schema, such as info related to customers or product catalogs, or even financial records.
- **Semi-structured data:** The semi-structured data refers to data that possesses some organization properties but does not strictly adhere to a schema. This includes XML files and JSON data, even NoSQL databases [15], with information arranged but flexible.

- **Unstructured data:** Major portions of big data are unstructured data. These are data that have no pre-defined structure or format, including videos, images, social media posts, emails, sensor data, etc. It is very hard to analyze unstructured data, but it does contain some real value insights once processed using advanced analytics techniques.

Each of these types poses different challenges for data processing, storage, and analysis, necessitating specialized tools and techniques to extract meaningful insights.

### 10.1.4 Big data analytics

Big data analytics, in simple words, is the process of analyzing large and complex datasets [16], [17] to identify unknown patterns, trends, or correlations that can lead to informed decision making. Advanced algorithms and computational methods [18], [19] applied for the identification of trends, making predictions about future outcomes, and performance optimization allow various companies of almost every industry to optimize operations, better serve customers with better customer experience, and get an edge.

Big data analytics is further subdivided into three types that are clearly mentioned in figure 10.2.

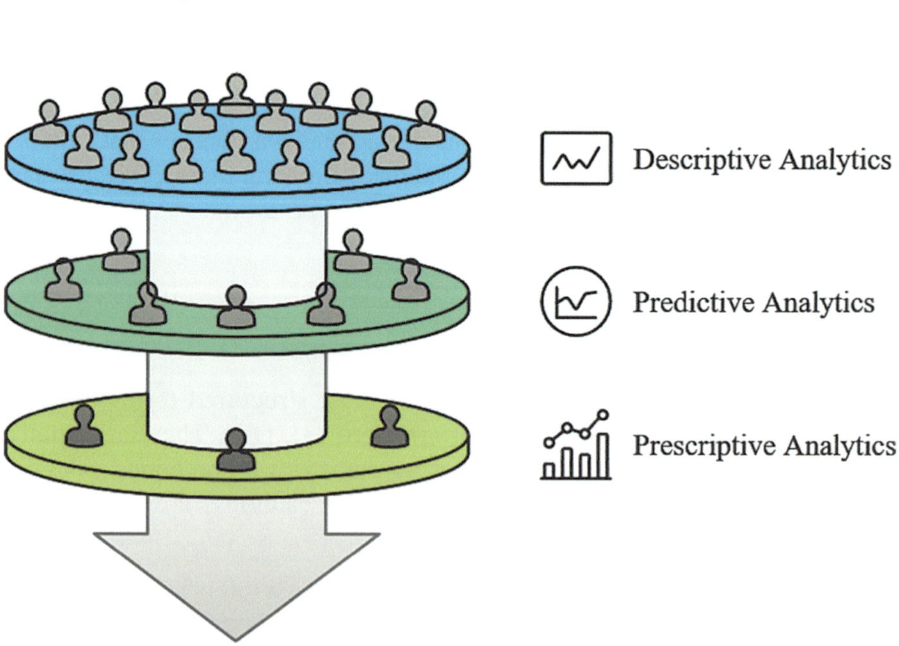


**Figure 10.2** Big data analytics process.

- **Descriptive analytics:** It draws inferences about summarizing historical data to know what happened in the past. Such analytics would typically generate reports and dashboards.
- **Predictive analytics:** Historical data, along with machine learning algorithms and statistical models, predict future events [20]. For example, it can predict which of its customers may be acting in a particular manner, which equipment might fail, or what the financial trend might be.
- **Prescriptive analytics:** This is beyond the ability to predict outcomes but, rather, prescribes what should be done to optimize business processes or operations [21]. Prescriptive analytics usually involve simulation and optimization models to determine the best course of action.

### 10.1.5 Big data analytics techniques

Big data analytics techniques to handle and analyze large datasets have been:

- Data mining: The process of unearthing patterns, relationships, and trends in large datasets. Data mining encompasses clustering, classification, and association rule mining, among other techniques, for the extraction of hidden information [10], [22].
- Machine learning: An area of artificial intelligence that involves using statistical techniques to enable machines to learn from data without explicit programming. Some of the major machine learning techniques used in big data analytics include supervised learning, unsupervised learning, and reinforcement learning, which make predictions or decisions based on historical data.
- NLP [23] is the ability of computers to recognize, interpret, and understand natural languages. It's being increasingly applied in big data analytics for sentiments, automating customer service, or delving into unstructured text data from social media and reviews.
- Sentiment analysis: This technique is applied to detect the emotions while describing opinions through text data such as customer feedbacks, social media posts, and reviews on the products. It presents a glimpse of how customers are feeling about a product or brand.
- Time series analysis: A methodology that is applied to analyze data points that represent temporal data to identify trends, cycles, or seasonal variations. Time series analysis [24] is generally applied in financial forecasting, weather forecasting, or demand forecasting.

### 10.1.6 Data tools and technologies

Big data requires highly sophisticated tools and technologies that can efficiently deal with the storage, processing, and analysis of immense datasets. Among the key tools and technologies used in big data analytics are as follows:

- **Apache Hadoop:** It is an open-source framework that facilitates the distributed storage and processing of large datasets [10]. It has been widely used for the purpose of batch processing of big data on clusters of computers.
- **Apache spark:** This is one of the fastest and easiest tools for real-time processing of big datasets. Spark provides in-memory computation abilities, which are useful for applications requiring fast iteration during processing.
- **NoSQL databases:** Although relational databases, for instance, MongoDB, Cassandra, and Couchbase, are tailored for large-scale handling of unstructured data, they allow flexibility in the storage of large amounts of data without respect to a fixed schema.
- **Data lakes:** Data lakes [25] are immense repositories that are designed to store vast amounts of raw data in its raw state. In contrast, from the data warehouse repository, data is stored a structured format and has data lakes storing structured as well as unstructured data.
- **Data Warehouses:** Unlike data lakes, which store messy or raw data, data warehouses store structured data that has been cleaned, organized, and integrated. So, it is thus optimized for query performance and best used for generating business intelligence reports and dashboards.

## 10.2 Computational Intelligence (CI) in Big Data Analytics

Given the ever-increasing volume and complexity of data, CI has emerged as an all-important add-on for the improvement of Big Data Analytics. CI provides techniques, including machine learning, neural networks, fuzzy logic, and evolutionary algorithms [7], [14], [16], which empower systems to learn from data and to make real-time decisions. Cybersecurity greatly depends on CI [5], [12], [26], so that patterns and anomalies seen by traditional methods may not be missed.

### 10.2.1 Role of CI in enhancing big data analytics

CI fundamentally changes the nature of how organizations will deal with information concerning data. Instead of rule-based analytics, algorithms used in CI learn from data, thereby increasing their adaptability to dynamic cyber threats (figure 10.3).

- **Scalability:** Algorithms that characterize CI like ANNs and SVMs work efficiently in processing large volumes of data, and this makes it be suitable for scenarios where there is constant production of data, such as cloud computing or IoT [3], [7], [27].
- **Flexibility:** CI models learn from new data, which means that there would be real-time detection of emerging threats in the field of cybersecurity.
- **Automation and accuracy:** CI automates the detection of threats; therefore, the time it would take from a human perspective reduces,

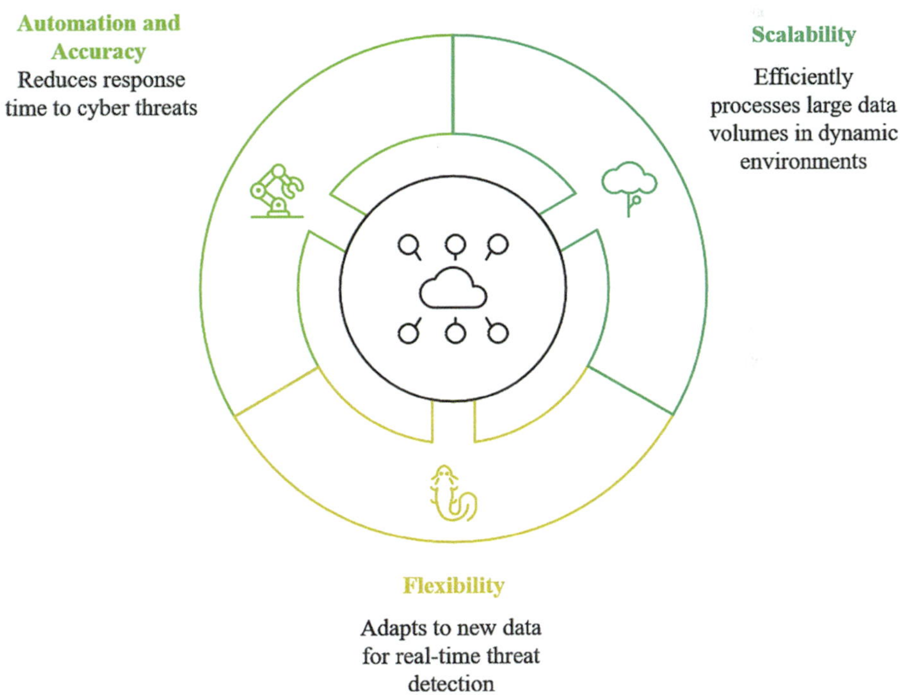


**Figure 10.3** Role of computational intelligence.

and the response time to attacks such as DDoS and malware infections increases.

### 10.2.2 Types of cyber threats addressed by CI

The adoption of CI improves the ability to find a wide range of cyber threats that fall into the following categories:

- **Insider threats:** The CI employs behavioral analytics and ANNs to establish anomalies in user behavior, which could be potential signs of an insider threat [3], [28].
- **Advanced persistent threats (APTs):** The ability to identify patterns in large datasets by CI models helps in detecting sophisticated attacks by pointing out deviations from normal behavior.
- **Ransomware and malware:** CI network traffic analysis detects and mitigates attacks by simulation with GANs [6], [29].
- **Phishing attacks:** NLP enables the detection of phishing emails; it is through analyzing the content of sent emails for suspicious elements [23], [30].

### 10.2.3 CI techniques for threat detection

Several CI methods enhance the detection and prevention of big data threats through the improvement of the following.

- **Artificial neural networks:** ANNs are astonishingly very efficient in pattern recognition; hence, anomaly detection on big data basically aids intrusion detection systems.
- **Fuzzy logic:** Fuzzy logic deals with uncertainty related to data, helping in ambiguous classifications of potential threats and adaptive defense mechanisms [31].
- **Genetic algorithms:** These algorithms optimize the feature selection for threat detection and adapt by improving evolving defense strategies [14].
- **Swarm intelligence:** PSO or Particle Swarm Optimization is the technique based on swarm intelligence used to optimize the network performance and predict some threats like a natural system [10], [11].

- **Support vector machines (SVMs):** SVMs are capable of classifying data to detect spam and malware by coming up with effective analysis of network traffic in intrusion detection systems (IDS) [27], [32].

## 10.3 Challenges in using CI for Threat Detection

While CI enhances the efficiency of cyber threat detection, several challenges degrade the effectiveness of using CI. This section delineates some significant hurdles to the successful implementation of CI models in real-world cybersecurity environments.

### 10.3.1 High dimensionality of data

Managing high-dimensional data is a tremendous challenge. The current networks generate much data most of the time with hundreds or thousands of attributes. More features generate more computing requirements which make it difficult for CI models to extract relevant insights efficiently. Techniques like PCA [9], [33] can reduce feature sizes but lose the relevant information simultaneously.

### 10.3.2 Data quality and noise

Quality of Input Data: The effectiveness of CI will thus depend on the quality of data [1], [31]. Noisy, incomplete, or inconsistent real-world data may degrade performance in models. It is also difficult to design preprocessing steps, such as cleaning and normalization, for large-scale datasets. Fuzzy logic can be useful to deal with uncertainty, but it should be calibrated carefully.

### 10.3.3 Scalability and real-time processing

With increased data volumes, scalability becomes an issue in CI systems. Most models would require copious resources since models like deep learning networks need to process real-time data. Distributed computing frameworks like Apache Hadoop and Spark can be used as scaling assistance; however, making this low latency is hard, especially with IoT data [11–13].

### 10.3.4 Interpretability and explainability

Most CI models are not interpretable, making it extremely difficult to understand the decision-making process behind them. It becomes a “black box” that complicates threat detection because security [34] teams need to know why a threat was flagged. Research in XAI [35] is aimed at improving model transparency, and it is challenging to achieve this by balancing complexity with human readability.

### 10.3.5 Privacy and security issues

CI operates on such data, so it is a question of privacy. To collect personal information to identify threats is again at risk of breach or misuse. Complying with regulations like GDPR, therefore, becomes paramount [36]. It is still in very preliminary stages of research into privacy-preserving techniques such as differential privacy.

### 10.3.6 Resource constraints

CI models, especially deep learning models, require enormous computational power, making them infeasible to most smaller organizations [19], [22]. Real-time detection of updates strains resources even more. The existing studies are more toward developing models that utilize much fewer resources without compromising accuracy.

Although promising advantages are seen in CI for detecting cyber threats, several challenges still need to be addressed concerning high-dimensional data, scalability, and explainability to make it entirely workable. Research issues in explainable AI and privacy-preserving techniques that are currently underway hold the promise of making more robust, scalable, and transparent versions of CI-driven cybersecurity solutions.

## 10.4 Applications

CI has changed the face of many industries through the sophistication of processing huge amounts of data, identification of patterns, and being proactive in making the right decisions in real-time. Some of the sectors mentioned in figure 10.4, where CI is creating a difference are in the contexts explained as follows, but primarily for addressing cyber threats.

### 10.4.1 Healthcare analytics

In the health sector, CI is particularly important in searching intricate data sources such as EHRs, medical images, or genetic data. Techniques applied to the area of predictive diagnostics may include ANNs [3]. They recognize early disease indicators, including conditions like cancer or cardiovascular diseases, by processing large volumes of patient data. For instance, ANN-based autoencoders can recognize possible insider threats in healthcare systems through learning and analysis of users' behavior and anomalies [2], [18].

Further, CI analytics provides security since it protects the sensitive health information from cyberattacks. It is through CI that hospitals and healthcare systems track attempted unauthorized access that leads to the breach of their patients' information.

### 10.4.2 Financial analytics

Almost all financial institutions are opting for CI models when it comes to fraud detection, risk management, and predictive analytics. SVMs and ANNs are deployed in order to identify anomalies related to transactional data, whereby suspicious activities are flagged in real-time [4], [20], [21]. Consequently, since this amounts to a preventive measure, cases of fraud in the form of credit card fraud, insider trading, and money laundering have plummeted dramatically.

CI further enables market forecasting by analyzing huge quantities of transactional financial data, stock prices, and economic indicators [17], [24]. Using complex pattern recognition capability, CI models can accurately forecast market trends to help in portfolio management at institutions by mitigating risks and making better investments.

### 10.4.3 Retail and e-commerce

In the retail area, CI is used in focused marketing, sales forecasting, and inventory control. Swarm Intelligence and Evolutionary Algorithms are utilized to optimize the supply chain by predicting the demand of the customers to reduce operational inefficiencies. For example, an ANN-based recommendation system by retailers like Amazon helps assess customer behavior and make a personalized product suggestion for the customer [15], [25].

Another area where CI is quite fundamental is the protection of e-commerce platforms against cyber threats. CI analyses huge volumes of online transactions and user interactions to identify fraudulent activities, secure payment systems, and prevent data breaches in real time.

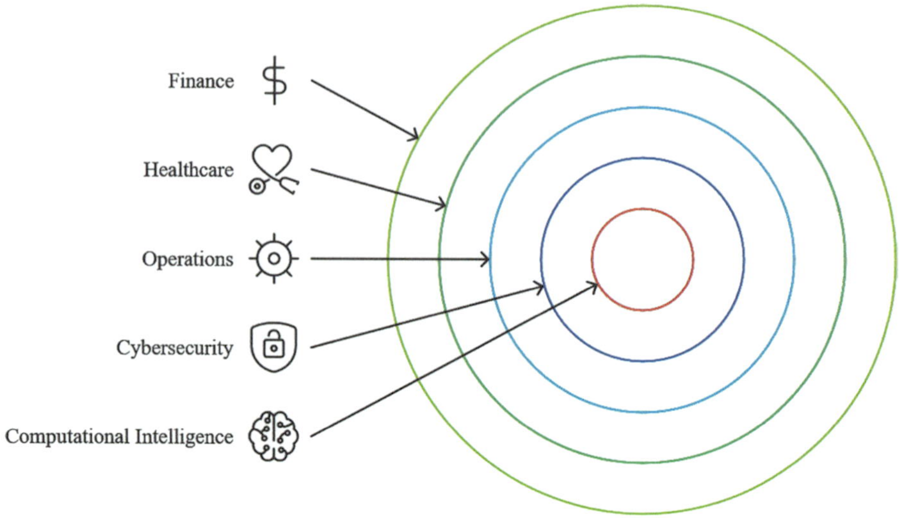


**Figure 10.4** Application of computational intelligence in various industries.

### 10.4.4 Smart manufacturing and industrial applications

CI techniques are applied in smart manufacturing for improving predictive maintenance and quality control through fuzzy logic and genetic algorithms; it analyze sensor data from industrial equipment to signal potential failures before a crash occurs, thus reducing downtime as well as operational costs and making the processes much more efficient [5], [31], [37].

CI also plays a significant role in ensuring the protection of industrial control systems from cyber-attacks. This is achieved through the continuous monitoring of network traffic and system logs by IDs, which are driven by CI in order to identify anomalies that could be indicative of malware infections, theft of SCADA systems, or improper access, among others [12], [26], [36].

Computational intelligence applied across industries reaffirms its applicability to both cybersecurity and operations. Across various industries, one sees how the use of computational intelligence can prove so versatile and effective in resolving problems both in cybersecurity and operations. Though CI is becoming increasingly adopted across healthcare, finance, and manufacturing, it keeps constantly moving toward innovation through processing vast scales of data, threat detection, and real-time decision-making to improve security and efficiency.

## 10.5 Case Studies and Real-World Applications

Computations are successfully implemented in various sectors to solve Big Data Analytics problem and cyber threat detection. Such case studies that change lives with the introduction of CI are as follows:

### 10.5.1 Personalized liver cancer risk prediction using the SMRF algorithm

The SMRF algorithm [6] was employed to predict the risk of developing liver cancer by scanning datasets using MRI, CT scans, and ultrasounds from medical imaging. The SMRF algorithm performed processing on large datasets in a distributed computing environment with an accuracy of 97.66% and with a significant decrease in computational time compared to previous algorithms.

This case illustrates the potential for CI-driven analytics to improve diagnosis, introduce interventions sooner, and more effectively tailor treatments for patients at higher risks.

### 10.5.2 Detection of financial transaction frauds using neural networks

A leading global bank implemented ANNs in its fraud detection systems [4], [20], [21], [27]. Based on real-time analysis of millions of daily transactions, the ANN model was able to identify unusual activity patterns and flag potential fraud. The system was able to reduce false positives by 60% and detect new fraud that the rules-based method had failed to catch earlier.

This case proves the use of CI for defeating highly advanced fraud schemes before they even reach financial institutions and ensure the security of transactions.

### 10.5.3 Predictive maintenance in smart manufacturing using fuzzy logic

A large vehicle manufacturing company was using fuzzy logic systems to predict machine failure by monitoring the sensor information it received from machines on the production lines. This allowed them to identify potential problems before they became real issues, cutting equipment downtime by 30% and millions in operational costs.

This is how CI techniques can refine industrial processes by preventing equipment failure and improving overall operational efficiency.

### 10.5.4 Artificial immune systems against FDI attacks on SCADA systems

The SCADA system protection with the use of artificial immune systems (AIS) in critical infrastructure due to the possible manipulation of sensor readings, even more seriously degradation of physical components, but required algorithms operating in real-time to successfully detect maliciously injected network packets and thereby avoid data manipulation and the deterioration of the integrity of SCADA systems [12], [26], [36].

This case demonstrates how CI techniques would protect the industrial control system from a cyberattack that may adversely impact the integrity of critical infrastructure.

It is through successful applications in healthcare, finance, manufacturing, and critical infrastructure [8], which confirm that CI is a leading technology for solving real-world problems. The efficiency of the systems in disease risk prediction, financial fraud detection, and cyberattack prevention speaks it about how CI can help overcome big data complexity and ensure cybersecurity [2–4], [18].

## 10.6 Future Directions for CI in Big Data Analytics

Based on further progress in technology, new ways of handling Big Data, and growing concerns over evolving cyber threats, the future promises to be transformative in CI detection of cyber threats. It is in these areas that CI matures further, and so will the revolution in its applications within the cybersecurity sphere.

### 10.6.1 Quantum computing and CI integration

One of the most promising developments in CI's near future is that of its integration with quantum computing. Quantum computers may be able to allow the processing power of CI algorithms to increase by orders of magnitude, thereby also providing for real-time analysis of large datasets that cannot be handled within a reasonable time by classical computers. QNNs are the new approach [14], [38] that can change network traffic or malicious patterns detection, thus significantly speeding up discovery and ensuring advanced capabilities in detecting threats like APT attacks.

Quantum-enhanced CI models will result in accurate predictions in dynamic environments, allow organizations to identify emerging cyber threats quickly, and offer protection when necessary.

### 10.6.2 Explainable AI (XAI) for improved transparency

Probably the biggest challenge to CI systems is transparency, often called the "black box" problem. Understanding how decisions are being made is difficult, given the complexity of modern CI models. In response, there is a high need for Explainable AI (XAI) primarily for industries such as finance and healthcare [9], where regulatory compliance often requires transparent and clear explanations of processes.

XAI techniques would allow cybersecurity professionals to determine why a CI system marks a particular activity as suspicious and which features of the data contributed to the decision [34], [35]. These models would not only increase the trustworthiness of CI systems but also make them much more actionable in real-world security operations.

### 10.6.3 Privacy-preserving CI models

Because organizations are increasingly relying on big data for the detection of cyber threats, data privacy and compliance issues with regulations, such as GDPR and HIPAA, are of great concern. Privacy-preserving CI models in the form of differential privacy and homomorphic encryption [19], [36] will allow organizations to analyze sensitive data without leaking privacy.

These models will enable CI systems to operate safely but in synchronization with the principles of data privacy, especially in such fields where personal or sensitive data are vulnerable. This would be more critical in ensuring threat detection systems analyze the data under strict controls of privacy.

### 10.6.4 Hybrid CI systems for adaptive cyber defense

The potential hybrid CI systems marry different CI techniques, such as ANNs, fuzzy logic, and genetic algorithms, to design more adaptable and stronger threat detection systems. Hybrid systems can better fit the complexity of contemporary cyber threats by taking advantage of the benefits realized from all constituent CI techniques.

For example, ANNs will be used for real-time pattern detection, fuzzy logic with imprecise data, and genetic algorithms [7], [11], [31] that optimize detection; therefore, this combined methodology will allow real-time adaptive flexibility to have better systems for cyber defense in the fight against a more diverse range of threats.

### 10.6.5 Real-time threat detection with edge computing

The data being generated at the edges of this network is rising exponentially with the growth of IoT and CPS. In future cyber threat detection, edge computing takes an important role [13], [23], [30] by letting CI models close the processing to where the data is sensed rather than sending it to some central computing site for processing, thus reducing the latency to real-time responses in its prediction.

Light CI models deployed at the edge point to the ability of organizations to efficiently recognize threats on IoT devices, smart grids, and ICS. This shift toward an edge-based CI will be particularly important in low-latency settings such as smart cities, autonomous vehicles, and critical infrastructure.

### 10.6.6 CI systems with human experts' end

Because interplays between CI-driven systems and human cybersecurity experts are going to be made much easier as CI systems further evolve, the latter will be allowed to exercise discretion during decision-making since they will be in charge of understanding context. Meanwhile, large-scale data analysis and pattern recognition will be taken care of by CI.

HITL systems, which intervene with a human analyst at various critical decision points, will allow more balanced cyber defense strategies and the continued learning and progression of CI models, coupled with human intuition and experience, particularly when an automated system does not make sense in tough threat scenarios [22], [24].

The future of CI in the detection of cyber threats is therefore excitingly new, with opportunities such as quantum computing and explainable AI, while there is a rise in the application of privacy-preserving models and edge computing. The latter will make CI more powerful and efficient, but also more transparent and flexible because CI must be able to better grasp new risks in real time to enable high-level decision-making to be kept informed by increasingly sophisticated cyber threats. It is through the incorporation of human expertise and other technological innovations that CI shall continue playing a central role in the shaping of the future of cybersecurity.

## 10.7 Conclusion

This chapter has discussed the significance of CI (computational intelligence) approaches in providing secure and safe system in the area of big data analytics. As data continues to grow rapidly, the complexity with which conventional security mechanisms are equipped to deal with on the cyber threat landscape will not perpetuate, nor will the generalization, heterogeneity, and volume of manifold cyber threats through different verticals. Coupled with artificial intelligence (AI), CI techniques have become a strong tool to analyze disparate data patterns, identify anomalies, and ensure data is protected against malicious behavior. The conversation also highlighted some specific types of cyber threats that can be effectively tackled by applying CI-enabled methodologies. In addition, the chapter highlights some associated challenges with using CI for security, such as scalability, adaptability, correctness, and adversarial attack. However, using CI for big data security has great promising prospects because it can be integrated into a variety of application domains like healthcare, financial, IoT ecosystems, and cloud infrastructures.

In terms of the future, the potential for CI with big data security is limitless and ever-expanding. One potential avenue for future research could be combining more than one CI and/or AI method to make predictions with higher accuracy and resistance to elaborate cyber threats. The other track is the adaptation of real-time technical frameworks for fast big data for on-time anomaly detection. In addition, as countries are devising new laws regarding data privacy and protection, explainable and interpretable CI models will become even more crucial to ensure transparency and trust in security-based decisions, especially in sensitive sectors like healthcare and finance. A second research direction will be to improve the robustness of CI models against adversarial attacks, as well as for resilience against shifting threat landscapes. Another trend is privacy-preserving intelligence, including federated learning and differential privacy, which protects information flow without compromising user privacy. And last but not least, CI-enabled security solutions can be pivoted to new and emerging domains like smart cities, autonomous vehicles, and quantum computing-based environments.

To sum up, although problems like these must be overcome, computational intelligence provides a new, evolutionary route to the cybersecurity of big data infrastructures. The constantly evolving progress with AI is resulting in the integration of smart methods with solid security systems, which will lead to next-level cyber defense systems that will be able to safeguard crucial data resources in the digital age.

## References


[1] R. Iqbal, F. Doctor, B. More, S. Mahmud, U. Yousuf, "Big data analytics: Computational intelligence techniques and application areas," Coventry University, 24 April 2018.

[2] R. Iqbal, F. Doctor, B. More, S. Mahmud, U. Yousuf, "Big Data analytics and Computational Intelligence for Cyber–Physical Systems: Recent trends and state of the art applications," Coventry University, 20 November 2017.

[3] M. Gomes, J. Kovaleski, R. Pagani, V. L. da Silva, T. C. D. S. Pasquini, "Transforming healthcare with big data analytics: technologies, techniques and prospects," Journal of Medical Engineering & Technology, 19 July 2022.

[4] A. C. Ikegwu, H. F. Nweke, C. V. Anikwe, "Recent trends in computational intelligence for educational big data analysis," Iran Journal of Computer Science, vol. 7, pp. 103–129, 7 September 2023.

[5] A. Gupta, "Big Data analysis using Computational Intelligence and Hadoop: A study," 2nd International Conference on Computing for Sustainable Global Development (INDIACom), New Delhi, India, pp. 1397-1401, 2015.

[6] A. Jain, A. Nadeem, H. M. Altoukhi, S. S. Jamal, H. K. Atiglah, H. Elwahsh, "Personalized Liver Cancer Risk Prediction Using Big Data Analytics Techniques with Image Processing Segmentation," Computational Intelligence and Neuroscience, 28 March 2022.

[7] M. G. Kibria, K. Nguyen, G. P. Villardi, O. Zhao, K. Ishizu, F. Kojima, "Big Data Analytics, Machine Learning, and Artificial Intelligence in Next-Generation Wireless Networks," IEEE Access, vol. 6, pp. 32328-32338, 2018.

[8] Sanjay Deshmukh, S. Sumeet, "Big Data Analytics Using Public Cloud Infrastructure: Use Cases and Cost Economics," International Conference on Computational Intelligence, 1 December 2015.

[9] B. A. Ojokoh, O. W. Samuel, O. M. Omisore, O. A. Sarumi, P. A. Idowu, E. R. Chimusa, A. Darwish, A. F. Adekoya, F. A. Katsriku, "Big data, analytics and artificial intelligence for sustainability," Federal University of Technology Akure, 18 September 2020.

[10] S. Sharma, V. K. Gahlawat, K. Rahul, R. S. Mor, M. Malik, "Sustainable Innovations in the Food Industry through Artificial Intelligence and Big Data Analytics," Logistics, vol. 5, no. 4, p. 66, 27 September 2021.

[11] A. M. Rahmani, E. Azhir, S. Ali, M. Mohammadi, O. H. Ahmed, M. Y. Ghafour, S. H. Ahmed, M. Hosseinzadeh, "Artificial intelligence

approaches and mechanisms for big data analytics: a systematic study," 14 April 2021.

[12] N. Mehta, S. Shukla, "Pandemic Analytics: How Countries are Leveraging Big Data Analytics and Artificial Intelligence to Fight COVID-19?," SN Computer Science, vol. 3, p. 54, 9 November 2021.

[13] N. N. Misra, Y. Dixit, A. Al-Mallahi, M. S. Bhullar, R. Upadhyay, A. Martynenko, "IoT, Big Data, and Artificial Intelligence in Agriculture and Food Industry," IEEE Internet of Things Journal, vol. 9, no. 9, pp. 6305-6324, 1 May 2022.

[14] S. Sivakumar, "Computational Intelligence Based Electronic Healthcare Data Analytics Using Feature Selection with Classification by Deep Learning Architecture," International Journal on Future Revolution in Computer Science and Medicine, 15 September 2022.

[15] S. R. Krishna, K. Rathor, J. Ranga, A. Soni, S. D., A. K. N., "Artificial Intelligence Integrated with Big Data Analytics for Enhanced Marketing," International Conference on Inventive Computation Technologies (ICICT), Lalitpur, Nepal, pp. 1073-1077, 2023.

[16] Ankit Verma, "Big Data Analytics Using Machine Learning Techniques for Prediction on Datasets," Computational Intelligence and Machine Learning, 14 April 2023

[17] L. Hui, P. Geczy, H. Lai, J. Gobert, S. J. H. Yang, H. Ogata, J. Baltes, R. Guerra, P. Li, C. C. Tsai, "Challenges and Future Directions of Big Data and Artificial Intelligence in Education," Frontiers in Psychology, vol. 11, 2020.

[18] D. Venkata Siva Reddy, R. Vasanth Kumar Mehta, "Cloud based computational intelligence approaches to machine learning and big data analytics: literature survey," 1 March 2018.

[19] X. Wu, W. Li, H. Tu, "Big data and artificial intelligence in cancer research," Review, vol. 10, no. 2, pp. 147–160, February 2024.

[20] D. Teh, T. Rana, "The Use of Internet of Things, Big Data Analytics and Artificial Intelligence for Attaining UN's SDGs," in *Handbook of Big Data and Analytics in Accounting and Auditing*, Springer, Singapore, 4 February 2023.

[21] S. Arora, S. R. Thota, "Using Artificial Intelligence with Big Data Analytics for Targeted Marketing Campaigns," IJARSCT, vol. 4, no. 3, June 2024.

[22] P. Balthazar, P. Harri, A. Prater, N. M. Safdar, "Protecting Your Patients' Interests in the Era of Big Data, Artificial Intelligence, and Predictive Analytics," Emory University School of Medicine, 1 March 2018.

[23] C. T. Chen, S. C. Chen, A. Khan, M. K. Lim, M. L. Tseng, "Antecedents of big data analytics and artificial intelligence adoption on operational performance: the ChatGPT platform," Industrial Management & Data Systems, vol. 124, no. 7, pp. 2388–2413, 6 June 2024.
[24] I. E. Agbehadji, B. O. Awuzie, A. B. Ngowi, R. C. Millham, "Review of Big Data Analytics, Artificial Intelligence and Nature-Inspired Computing Models towards Accurate Detection of COVID-19 Pandemic Cases and Contact Tracing," International Journal of Environmental Research and Public Health, vol. 17, no. 15, p. 5330, 24 July 2020.
[25] L. Ma, R. Chang, "How big data analytics and artificial intelligence facilitate digital supply chain transformation: the role of integration and agility," Management Decision, 2024.
[26] B. Hu, Zhenhan Zhang, "Evaluation of Big Data Analytics and Cognitive Computing in Smart Health Systems," Journal of Commercial Biotechnology, 21 August 2022.
[27] H. M. Shah, B. B. Gardas, V. S. Narwane, H. S. Mehta, "The contemporary state of big data analytics and artificial intelligence towards intelligent supply chain risk management: a comprehensive review," Kybernetes, vol. 52, no. 5, pp. 1643–1697, 5 May 2023.
[28] H. Zhang, M. Song, H. He, "Achieving the Success of Sustainability Development Projects through Big Data Analytics and Artificial Intelligence Capability," Sustainability, vol. 12, no. 3, p. 949, 28 January 2020.
[29] A. A. A. Gad-Elrab, "Modern Business Intelligence: Big Data Analytics and Artificial Intelligence for Creating the Data-Driven Value," IntechOpen, 19 May 2021.
[30] A. Ghimire, S. Thapa, A. K. Jha, S. Adhikari, A. Kumar, "Accelerating Business Growth with Big Data and Artificial Intelligence," Fourth International Conference on I-SMAC, Palladam, India, pp. 441-448, 2020.
[31] A. Rosenfeld, D. Benrimoh, C. Armstrong, N. Mirchi, T. Langlois-Therrien, C. Rollins, M. Tanguay-Sela, J. Mehltretter, R. Fratila, S. Israel, E. Snook, K. Perlman, A. Kleinerman, B. Saab, M. Thoburn, C. Gabbay, A. Yaniv-Rosenfeld, "Big Data analytics and artificial intelligence in mental healthcare," Bar-Ilan University, 19 March 2021.
[32] R. C. Carlos, C. E. Kahn, S. Halabi, "Data Science: Big Data, Machine Learning, and Artificial Intelligence," Review, vol. 15, no. 3, pp. 497-498, March 2018.

[33] W. A. Chaovalitwongse, Y. Yuan, Q. Zhang, "Special issue: Innovative applications of big data and artificial intelligence," Frontiers of Engineering Management, vol. 9, pp. 517–519, December 2022.

[34] D. Dai, Sahar Boroomand, "A Review of Artificial Intelligence to Enhance the Security of Big Data Systems: State-of-Art, Methodologies, Applications, and Challenges," Archives of Computational Methods in Engineering, 12 July 2021.

[35] Z. H. Munim, M. Dushenko, V. J. Jimenez, M. H. Shakil, M. Imset, "Big data and artificial intelligence in the maritime industry: a bibliometric review and future research directions," 12 July 2020.

[36] H. Kamyab, T. Khademi, S. Chelliapan, M. SaberiKamarposhti, S. Rezania, M. Yusuf, M. Farajnezhad, M. Abbas, B. Hun Jeon, Y. Ahn, "The latest innovative avenues for the utilization of artificial Intelligence and big data analytics in water resource management," Results in Engineering, vol. 20, p. 101-566, 2023.

[37] W. A. Chaovalitwongse, Y. Yuan, Q. Zhang, "Special issue: Innovative applications of big data and artificial intelligence," Frontiers of Engineering Management, vol. 9, pp. 517–519, December 2022.

[38] S. Zabala-Vargas, M. Jaimes-Quintanilla, M. H. Jimenez-Barrera, "Big Data, Data Science, and Artificial Intelligence for Project Management in the Architecture, Engineering, and Construction Industry: A Systematic Review," Buildings, vol. 13, no. 12, p. 2944, 25 November 2023.

# 11

# ML-Driven Approaches to Improve Cybersecurity in Financial Transactions

**Shiv Ashish Dhoundiyal[1], Vinesh[2], Asheesh Kumar[3], Sushil Chandra Dimri[4] and Rekha Deoli[5]**

[1]Department of Computer Science and Engineering, Graphic Era Deemed to be University, India
[2]IIMT College of Engineering, India
[3]GNIOT Institute of Professional Studies, India
[4]Department of Computer Science and Engineering, Graphic Era Deemed to be University, India
[5]Department of Computer Science and Engineering Graphic Era Hill University, India

**Email:** shivashish1234@gmail.com; ranavini700@gmail.com; ashishkumartyagimba@gmail.com; dimri.sushil2@gmail.com; deolirekha@gmail.com

## Abstract

Cyber risks pose substantial problems in preserving data integrity, privacy, and security in the continuously evolving context of financial transactions. Specifically, this study investigates how logistic regression, support vector machine, decision tree, and random forest might be applied to improve the cybersecurity of financial systems. The algorithms are tested to see how well they can detect suspicious activity, spot fraud, and reduce cyber threats. The paper proves that these models can attain high recall, precision, and accuracy through extensive experimentation on real-world financial datasets. In light of these results, it is clear that ML-driven solutions hold great promise for improving financial cybersecurity by offering effective, scalable, and resilient defenses against evolving cyber threats.



DOI: 10.1201/9788743812739-11

## 11.1 Introduction

Despite the innumerable resources and materials put into the modern world's security systems, the proper security of Taiwan's financial institutions faces a lot of challenges as listed above. Crime and aggressive malicious activity continuously evolve as the cyber world changes and adapts to the modern world's reality. The resilience of Taiwan's financial institutions has, in the past, led many amateurs and experts to believe that there are very few challenges standing in the way of their success [1, 2]. This has, however, been proven wrong as banking institutions around the world have begun witnessing money laundering, fraud, and payment-linked financial crime at an alarming rate, along with an increase in corporate balance and an increase in political tension, coupled with a shift in power dynamics. Cybercrime is no longer centered within borders, which only creates stronger implications and consequences for all financial institutions, banks, and dealings around the world. Taiwan, in recent years, has seen more aggressive action being taken against Chinese interference, which has led to many countries distancing themselves and beginning to take notice of how strong an entity Taiwan actually is, but how oppressed it has been by the domination of China [3-5]. We hope to learn more about how well these three algorithms work in identifying and reducing cyber risks to financial systems by conducting this study. The statistical model known as logistic regression is used in binary classification applications due to its simplicity and interpretability. In the case of mining decision rules through data, the decision tree is unmatched in terms of hierarchy. Nonetheless, Random Forest, a combination strategy that relies on numerous decision trees, can guarantee superior accuracy and robustness [6-8].

This study determines the extent to which these models address significant cybersecurity issues, such as fraud detection and anomaly detection, through extensive experiments. The findings can be used to develop more powerful and secure financial systems and bring clarity to the practical application of ML approaches to address financial cybersecurity [9, 10].

## 11.2 Methodology Used

### 11.2.1 Raw data collection

The data collection process involves the collection of information through data processing of collections using a very broad spectrum of sources. To analyze the events in the past and a trend, the collection of data is required. The dataset of the fraud detection can be obtained through Kaggle. It is a simulation of the mobile money transactions with the aim of gaining a better insight into the operation of mobile money transaction in the real world, and the additional objective of studying fraudulent behaviors in academia [11-14]. It presents multiple types of transactions, such as cash-in, cash-out, debit, payment, and transfer, within a simulated timeframe of 30 days, which offer a realistic hands-on environment to study fraud detection techniques [15-17].

### 11.2.2 Data pre-processing

Pre-processing data enables analysts and data scientists to use machine learning algorithms to gain insights or predict future events. This research process involves identifying and correcting missing values. Collecting data points for each dataset entry can be challenging. Missing data may consist of empty cells, null values, or question marks. The research dataset has no missing values [18-22].

Figure 11.1 represents the sample dataset used in the study, the dataset process using Python version 3.

Figure 11.2 represents different transaction types (payment, transfer, debit, cash, etc.) and their corresponding frequency.

Figure 11.3 represents the number of transactions in terms of the number of counts per month.

### 11.2.3 Train and testing dataset

Partitioning of the dataset into testing and testing datasets. The 2200-dataset has been partitioned as follows: training dataset (1760 data) comprises 80% of the set, while testing dataset (440 data) comprises the remaining 20% of the set.

| | | step | amount | nameOrig | oldbalanceOrg | newbalanceOrig | nameDest | oldbalanceDest |
|---|---|---|---|---|---|---|---|---|
| type | isFraud | | | | | | | |
| CASH_IN | 0 | 1399284 | 1399284 | 1399284 | 1399284 | 1399284 | 1399284 | 1399284 |
| CASH_OUT | 0 | 2233384 | 2233384 | 2233384 | 2233384 | 2233384 | 2233384 | 2233384 |
| | 1 | 4116 | 4116 | 4116 | 4116 | 4116 | 4116 | 4116 |
| DEBIT | 0 | 41432 | 41432 | 41432 | 41432 | 41432 | 41432 | 41432 |
| PAYMENT | 0 | 2151495 | 2151495 | 2151495 | 2151495 | 2151495 | 2151495 | 2151495 |
| TRANSFER | 0 | 528812 | 528812 | 528812 | 528812 | 528812 | 528812 | 528812 |
| | 1 | 4097 | 4097 | 4097 | 4097 | 4097 | 4097 | 4097 |

**Figure 11.1** Dataset sample used.

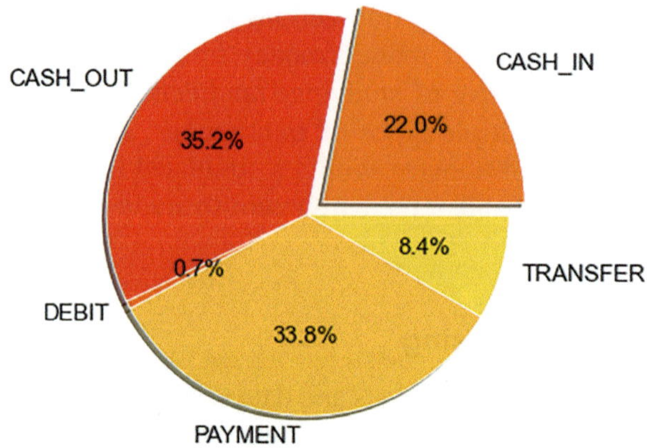


**Figure 11.2** Percentage count of each type of transaction in the dataset.

### 11.2.4 Fitting the model

Model parameters are fitted to improve accuracy. Machine learning models are created by algorithms processing target variable data. Model accuracy is determined by comparing outputs to objective variable values. A machine learning model's capacity to generalize data like its training is termed "model

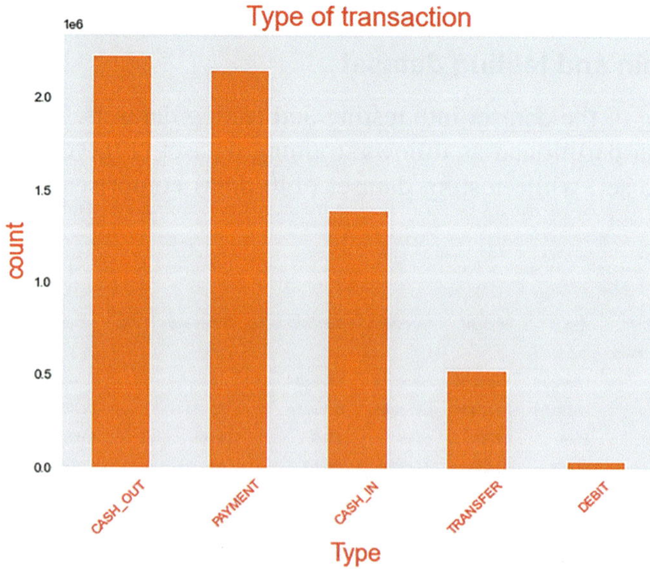


**Figure 11.3** Different types of transaction in the dataset.

fitting." A good model fits unknown inputs and accurately approximates the result [23-26].

### 11.2.5 Predicting the model

- Logistic regression
- Support vector machine
- Decision tree
- Random forest

The term "prediction" in the context of estimating the probability of a particular outcome denotes the result generated by an algorithm subsequent to its training on a prior dataset and application to novel data. Implementing the predict () method with the test feature dataset to forecast the model. The output consists of a collection of predicted values [19].

Figure 11.4 represents the framework of the proposed model for fraud prediction in cybersecurity of financial systems. Relevant features are extracted using both traditional signal processing techniques and deep feature extraction.

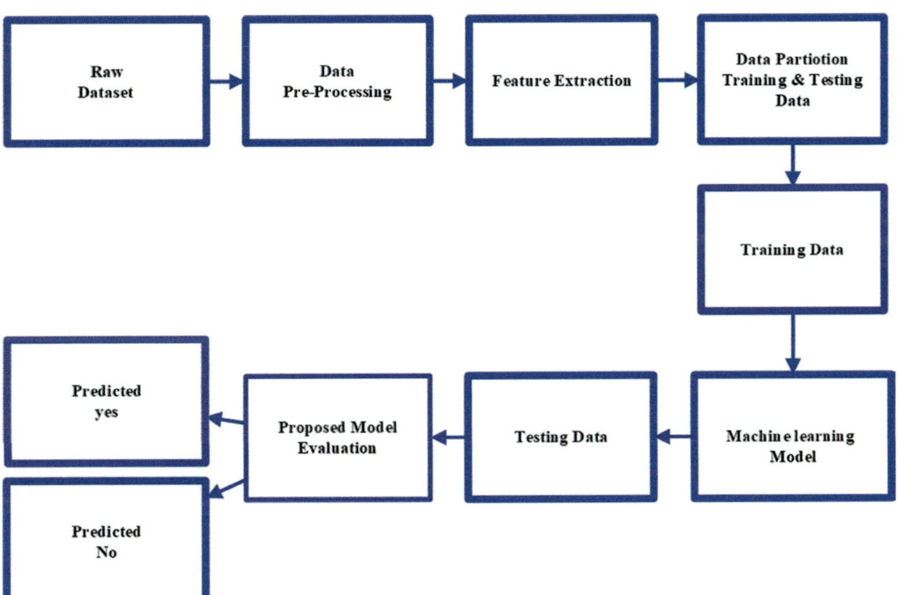


**Figure 11.4** Framework for the proposed optimized RF model.

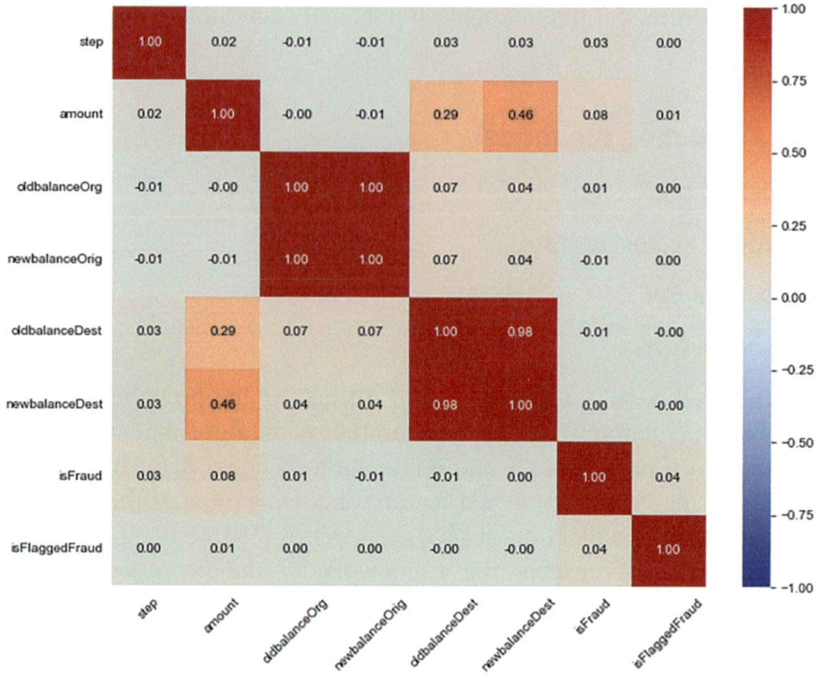


**Figure 11.5** Confusion matrix for the dataset used.

### 11.2.6 Confusion matrix

It furnishes a matrix displaying the occurrence of accurate negatives, accurate positives, and inaccurate positives.

Figure 11.5 is a confusion matrix depicting the classification performance of the proposed model on the fraud detection dataset. The matrix illustrates the distribution of true positive, true negative, false positive, and false negative predictions, providing a detailed insight into the proposed model's predictive accuracy across different classes.

### 11.2.7 Classification report

Use this statistic to see how well a classification algorithm does at making predictions. A model's output includes three measures: F1 score, recall, and precision.

**Accuracy:** Accuracy measures how often a machine learning model predicts the outcome. Accuracy is exact predictions divided by total forecasts [27, 28].

**Precision:** Classifier skill in identifying accurate positive predictions. Genuine positives divided by true and false positives [29, 30].

**Recall:** Classifier sensitivity detects all confusion matrix positives. Calculate each category's accurate positive prediction % of correct positive and negative predictions. Recall measures the positive detection rate. False negative [31, 32].

**F1 score**: This metric is a different approach to accuracy for evaluating machine learning models; instead of looking at the model as a whole, it focuses on how well it predicts outcomes within specific classes [33-35].

## 11.3 Results and Discussion

Evaluation of performance was conducted utilizing parameters including F1 score, recall, precision, and F1 score. The use of these indicators, which are widely employed by most provincial systems for the identification of heart illness, enables a justifiable comparison between our approach and provincial systems. The definitions of these measurements are as follows:

$$Accuracy = \frac{TP + TN}{Total\,Samples} \tag{11.1}$$

$$Recall\left(Sensitivity\right) = \frac{TP}{TP + FN} \tag{11.2}$$

$$Precison\left(Specificity\right) = \frac{TP}{TP + FP} \tag{11.3}$$

$$F1_{Score} = 2 * \frac{Precison * Recall}{Precison + Recall} \tag{11.4}$$

Here,

TP is true positive,

TN is true negative,

FP is false positive, and

FN is false negative.

Table 11.1 shows the performance comparison of various machine learning models applied to the fraud detection task. The models are evaluated

**Table 11.1** Comparison of the performance of various ML models.

| Models | Accuracy (%) | Precision (%) | Recall (%) | F1 score (%) |
|---|---|---|---|---|
| LR model | 78.49 | 86.39 | 67.13 | 75.55 |
| Optimized RF model | 87.62 | 87.65 | 87.30 | 87.47 |
| SVM model | 84.58 | 84.04 | 85.00 | 84.52 |
| DT model | 85.02 | 84.29 | 85.74 | 85.01 |

based on key performance metrics such as accuracy, precision, recall, and F1 score. The results demonstrate the relative effectiveness of each model in capturing discriminative features from EEG signals.

Figure 11.6 Accuracy comparison of different machine learning models employed for fraud detection classification. The figure illustrates the performance of each model in terms of classification accuracy, the proposed optimized RF model outperforms the other ML models in terms of accuracy.

Figure 11.7 shows a precision comparison of various machine learning models used for fraud detection classification. The figure highlights the ability of each model to correctly identify positive instances, with the proposed optimized RF model achieving the highest precision among all evaluated approaches.

Figure 11.8 shows a recall performance of various machine learning models applied to fraud detection dataset classification. The figure illustrates

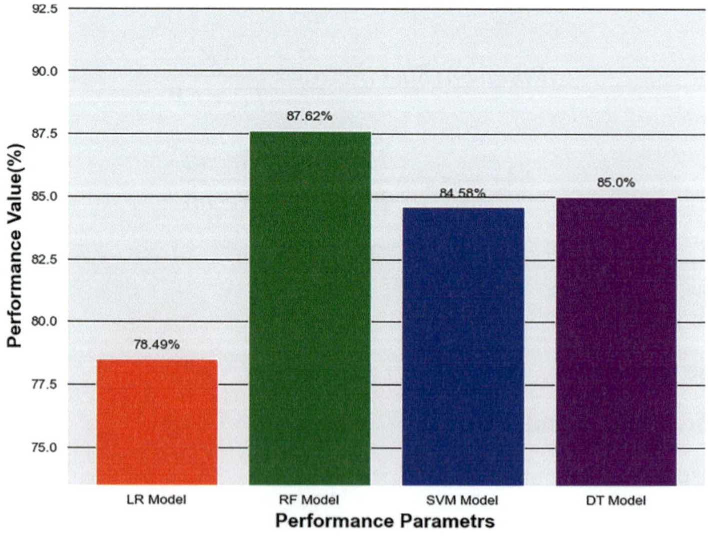


**Figure 11.6** Accuracy performance of various ML models.

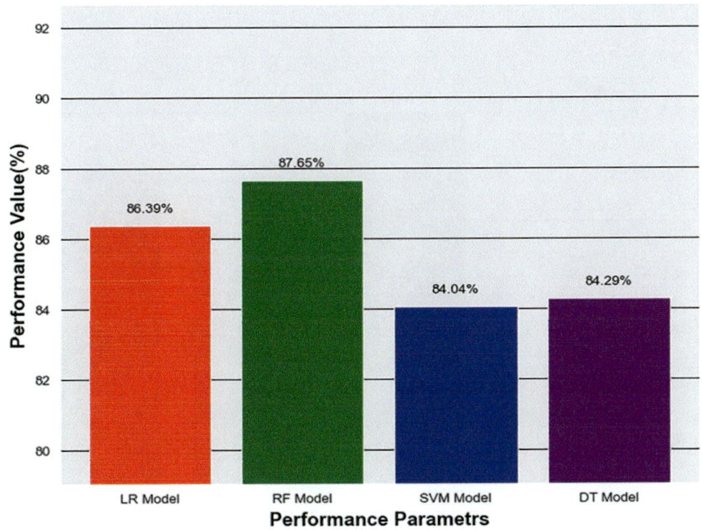


**Figure11.7** Precision performance of ML models.

each model's ability to correctly identify true positive instances, with the proposed optimized RF model demonstrating superior recall compared to the other ML approaches.

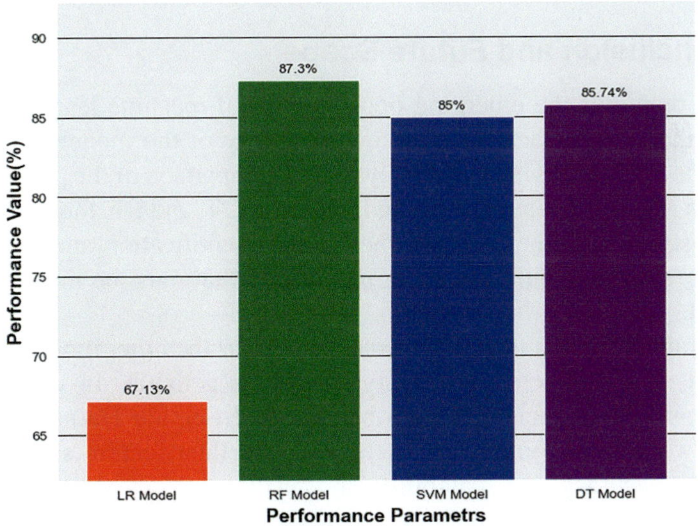


**Figure 11.8** Recall performance of ML models.

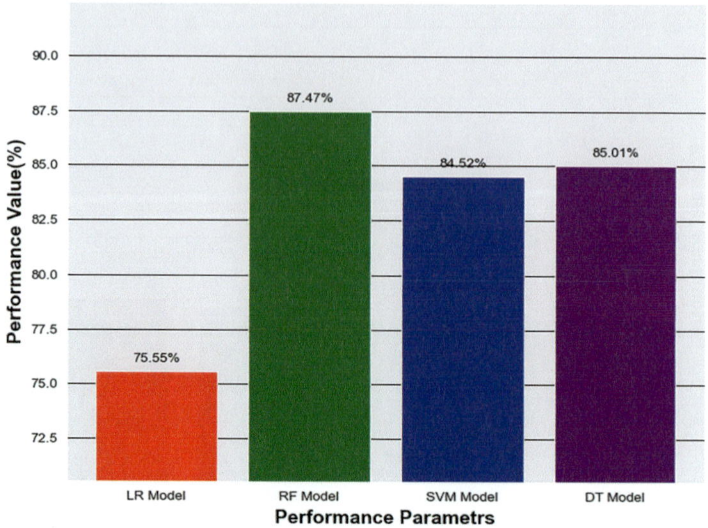


**Figure 11.9** F1 score of ML models.

Figure 11.9 shows F1 score comparison of different machine learning models employed for fraud detection classification. The F1 score, which balances precision and recall, reflects the overall effectiveness of each model. The F1 score for random forest performs better than the other approaches.

## 11.4 Conclusion and Future Scope

The study mentions the emerging opportunities of machine learning (ML)-driven methods to better assure the cybersecurity of the monetary transactions. As the study focuses on the analysis and comparison of the performance of multiple ML algorithms, including LR, SVM, DT, and RF, the paper highlights the strength of these approaches to the identification and mitigation of cyber threats by identifying exact parameters that must be measured and minimized.

The random forest technique exceeded by far the other models that are considered. It can effectively identify fraudulent activities in variable and high-dimensional financial data attributable to ensemble learning, built-in ability to work with complex datasets, and excellent feature selection. To enhance the accuracy and scalability, and resilience of cybersecurity risks, the concrete role of powerful ensemble methods becomes imperative as per our discovery. The advantages of the random forest that financial institutions can take are the ability to optimize the efficiency of fraud detection systems, minimize false positives, and reinforce their security against sophisticated cyberattacks. These findings indicate the growing sophistication of the risks involving the financial sector, posing the issue of pre-integrating ML-based approaches into the current designs of cybersecurity.

## Acknowledgments

We would like to express our sincere gratitude to the organizations and individuals who offered their help with regard to this research. We would like to sincerely thank the team, who provided vital technical support, considerate debates, and guidance throughout the whole research process. We would also like to extend special recognition to our advisors and colleagues who gave their best input and overall support.

## References

[1] Md, Al-Imran., Eftekhar, Hossain, Ayon., Md, Rashedul, Islam., Fuad, Mahmud., Sharmin, Akter., M., K., Alam., Md, Tarek, Hasan., Sadia, Afrin., Jannatul, Ferdous, Shorna., Md., Abdul, Aziz. "Transforming banking security: the role of deep learning in fraud detection systems." The American journal of engineering and technology, (2024). doi: 10.37547/tajet/volume06issue11-04.

[2] S., Bhuvaneswar., B., Avyay., Kondadi, Tejith., Ms., S, Kavitha. "A Supervised Ml Algorithm for Detecting and Predicting Fraud Credit Card Transactions." Deleted Journal, (2024). doi: 10.47392/irjaeh.2024.0349.

[3] Michal, Srokosz., Andrzej, Bobyk., Bogdan, Księżopolski., M., Wydra. "Machine-Learning-Based Scoring System for Antifraud CISIRTs in Banking Environment." Electronics, (2023). doi: 10.3390/electronics12010251.

[4] Shailendra, Mishra. "Exploring the Impact of AI-Based Cyber Security Financial Sector Management." Applied Sciences, (2023). doi: 10.3390/app13105875.

[5] Jane, K., Howard. "Cybersecurity and Fraud Detection in Financial Transactions." (2022). doi: 10.1007/978-3-030-94590-9_15.

[6] Ali, Alhchaimi. "Cloud-Based Transaction Fraud Detection: An In-depth Analysis of ML Algorithms." Wasit journal of computer and mathematics science, undefined (2024). doi: 10.31185/wjcms.253.

[7] B., Narsimha., Ch., V., Raghavendran., Pannangi, Rajyalakshmi., G., K., Reddy., M., S., Bhargavi., Pallavi, Naresh. "Cyber Defense in the Age

of Artificial Intelligence and Machine Learning for Financial Fraud Detection Application." International journal of electrical & electronics research, (2022). doi: 10.37391/ijeer.100206.

[8] Rajiv, Tulsyan,Pranjal, Shukla., Tushar, Singh., Anshul, Bhardwaj. "8. Cyber Security Threat Detection Using Machine Learning." Indian Scientific Journal of Research in Engineering and Management, (2024). doi: 10.55041/ijsrem37949.

[9] Iqra, Naseer. "Machine Learning Applications in Cyber Threat Intelligence: A Comprehensive Review." Deleted Journal, (2024). doi: 10.62019/abbdm. v3i2.85.

[10] C., L., Udeze, Idongesit, Efaemiode, Eteng., Ayei, E., Ibor. "Application of Machine Learning and Resampling Techniques to Credit Card Fraud Detection." Journal of Nigerian Society of Physical Sciences, (2022). doi: 10.46481/jnsps.2022.769

[11] Ngai, E. W. T., Hu, Y., Wong, Y. H., Chen, Y., & Sun, X. (2011). The application of data mining techniques in financial fraud detection: A classification framework and an academic review of literature. *Decision Support Systems*, 50(3), 559-569.

[12] Pacheco, F., & Krishnamoorthy, M. (2016). Financial fraud detection and data protection in big data. Journal of Banking and Finance, 12(4), 245-257.

[13] Conti, M., Dehghantanha, A., Franke, K., & Watson, S. (2018). Internet of Things security and forensics: Challenges and opportunities. *Future Generation Computer Systems, 78*, 544–546. https://doi.org/10.1016/j.future.2017.07.060.

[14] Raghavan, S., & Tripathi, A. (2018). Application of SVM and random forest for detecting online transaction frauds. International Journal of Data Mining and Emerging Technologies, 8(1), 30-36.

[15] R. Kashyap, "Big data analytics for smart cities: Trends and challenges," *IEEE Trans. Ind. Informat.*, vol. 20, no. 4, pp. 5071–5083, Apr. 2024.

[16] D. Singh, P. Sharma, M. A. Khan, and R. Bedi, "Artificial intelligence in cybersecurity: Threat detection and prevention," IEEE Trans. Inf. Forensics Security, vol. 19, pp. 456–469, Feb. 2024.

[17] S. Patil, R. Singh, and N. Kumar, "Reinforcement learning-based optimization for edge computing in IoT networks," IEEE Trans. Netw. Serv. Manag., vol. 20, no. 1, pp. 342–356, Jan. 2024.

[18] Y. Ren, Y. Xiao, Y. Zhou, Z. Zhang, and Z. Tian, "Cskg4apt: A cybersecurity knowledge graph for advanced persistent threat organization attribution," *IEEE Trans. Knowl. Data Eng.*, vol. 35, pp. 5695–5709, Aug. 2022.

[19] A. S. Rajawat, S. B. Goyal, P. Bedi, T. Jan, M. Whaiduzzaman, and M. Prasad, "Quantum machine learning for security assessment in the Internet of Medical Things (IoMT)," *Future Internet*, vol. 15, no. 8, p. 271, Aug. 2023.

[20] A. Piplai, A. Kotal, S. Mohseni, M. Gaur, S. Mittal, and A. Joshi, "Knowledge-enhanced neurosymbolic artificial intelligence for cybersecurity and privacy," *IEEE Internet Comput.*, vol. 27, no. 5, pp. 43–48, Sept.–Oct. 2023.

[21] R. Kour, R. Karim, P. Dersin, and N. Venkatesh, "Cybersecurity for Industry 5.0: Trends and gaps," *Front. Comput. Sci.*, vol. 6, p. 1434436, Mar. 2024.

[22] P. R. J. Trim and Y.-I. Lee, "Advances in cybersecurity: Challenges and solutions," *Appl. Sci.*, vol. 14, no. 10, p. 4300, May 2024.

[23] S. Ali and M. Mushtaq, "Publication pattern and research assessment of cybersecurity: A bibliometric study," in Bibliometrics – An Essential Methodological Tool for Research Projects. London, U.K.: IntechOpen, 2024.

[24] J. Crabb, C. Hundhausen, and A. Gebremedhin, "A critical review of cybersecurity education in the United States," in *Proc. 55th ACM Tech. Symp. Comput. Sci. Educ.*, Portland, OR, USA, Mar. 2024, pp. 241–247.

[25] J. Crabb and A. Gebremedhin, "Cybersecurity education and research: Experiences in training the next generation of cyber professionals," *Cyber Mag. Mil. Cyber Prof. Assoc.*, vol. 8, pp. 15–22, 2024.

[26] S. Sadeghi et al., "SoK: A comprehensive analysis and evaluation of Docker container attack and defense mechanisms," in *Proc. IEEE Symp. Secur. Privacy (S&P)*, San Francisco, CA, USA, May 2024, pp. 4573–4590.

[27] N. Papernot et al., "Phantom: Untargeted poisoning attacks on semi-supervised learning," in *Proc. 33rd USENIX Security Symp.*, Anaheim, CA, USA, Aug. 2024.

[28] H. Xiao et al., "LayerDBA: Circumventing similarity-based defenses in federated learning," in *Proc. 7th Deep Learn. Secur. Workshop (DLS)*, San Francisco, CA, USA, June 2024.

[29] Y. Wang et al., "Lost and found in speculation: Hybrid speculative vulnerability detection," in *Proc. IEEE/ACM Design Autom. Conf. (DAC)*, San Francisco, CA, USA, June 2024.

[30] L. Zhou et al., "AI-assisted threat intelligence for cloud-native environments," *IEEE Cloud Comput.*, vol. 11, no. 2, pp. 58–69, Mar.–Apr. 2025.

[31] H. Kheddar, Y. Himeur, and A. I. Awad, "Deep transfer learning for intrusion detection in industrial control networks: A comprehensive review," *J. Netw. Comput. Appl.*, vol. 220, Art. 103760, 2023.
[32] P. Wasnik and N. Chavhan, "A review paper on designing intelligent intrusion detection system using deep learning," in Proc. 2023 11th Int. Conf. Emerg. Trends Eng. Technol. – Signal Info. Process. (ICETET-SIP), Nagpur, India, 2023, pp. 1–6, doi: 10.1109/ICETET-SIP58143.2023.10151563.
[33] V. Sivagaminathan, M. Sharma, and S. K. Henge, "Intrusion detection systems for wireless sensor networks using computational intelligence techniques," Cybersecurity, vol. 6, Art. 27, 2023.
[34] F. Ullah, S. Ullah, G. Srivastava, and J. C.-W. Lin, "IDS-INT: Intrusion detection system using transformer-based transfer learning for imbalanced network traffic," Digit. Commun. Networks, vol. 10, pp. 190–204, 2024.
[35] H. Sadia, S. Farhan, Y. U. Haq, R. Sana, T. Mahmood, S. A. O. Bahaj, et al., "Intrusion detection system for wireless sensor networks: A machine learning based approach," IEEE Access, vol. 12, pp. 52565–52582, 2024.

# Index

# About the Editors

**Prof. (Dr.) Sushil Chandra Dimri**, M.Tech., PhD (Computer Science), currently working with Graphic Era deemed to be University as professor in Department of computer Science and engineering. His teaching industry experience is more than 26 years. Dr. Sushil Chandra Dimri authored 9 books on Mathematics and Computer Science i.e. Engineering Mathematics, Computer Based Numerical Techniques, Probability and Statistics, Graphical User Interface, Web Programming, Fundamental of Computers, Algorithms-Design and analysis, C- Programming and Mathematical Foundation of Computers. He published 60+ research papers in national and international journals. His areas of interest are Machine learning techniques, Designing algorithms and Resource Optimization. He is editor-reviewer for many reputed journals.

**Prof. (Dr.) Lata Nautiyal** is an accomplished computer science researcher and academician with extensive expertise in Software Engineering and Cyber Security. With over a decade of experience in academic teaching, research, supervision, and leadership, she has demonstrated exceptional ability to work both independently and as a collaborative team member. Her research interests focus on Cyber Security and Security Requirements Engineering, striving to innovate and enhance technology to better align with user needs and expectations. She owns 22+ Repute research papers in various National and International Journals along with Editor, reviewer in various international journals.

**Dr. Akshay Kumar** received the B.Sc. and M.Sc. degree in science from Chaudhary Charan Singh University, Meerut, India, in 2010 and 2012, and the Ph.D. degree major in Mathematics and minor in Computer science from G. B. Pant University of Agriculture and Technology, Pantnagar, India, in 2017. He is currently working as an Assistant Professor in the Department of Mathematics, Graphic Era Hill University, Dehradun in India. He has more than 8-year teaching experience. He is a regular Reviewer for more than 60 international journals including Elsevier, Springer, Emerald, MDPI, Cambridge University Press, World Scientific, River Publisher, GI Global Publisher and many other publishers. He has published 100 plus research papers in Communications in Statistics - Simulation and Computation., Elsevier, Emerald, inderscience, International Journal of Mathematical, Engineering and Management Sciences, IGI Global Publisher, MDPI, Cambridge Scientific Publishers, Springer nature, Multimedia Tools and Applications and many other national and international journals of repute and presented his works at national and international conferences. Also, he has published 7 books (edited) with international publishers like River Publisher, CRC Press Taylor & Francis, Elsevier, De- Gruyter and IGI Global Publisher. His fields of research are reliability theory, fuzzy reliability, signature reliability and applied Mathematics.

**Dr. Richa Indu** (Ph.D.) in Computer Science and Engineering from Graphic Era (Deemed to be University), Dehradun, INDIA) Richa Indu accomplished M.Tech (Hons) in Computer Science and Engineering from Uttarakhand Technical University, Dehradun. Furthermore, she has completed M.Sc. (Gold medalist) in Information Technology from Hemavati Nandan Bahuguna University, Srinagar (Garhwal), INDIA. She has made some noteworthy contributions to the field with 7 published research articles in reputed journals, 5 conference papers, and 2 book chapters. Her research interests encompass data sciences, programming languages, and algorithm design, reflecting her effort to advance the frontiers of technology.

For Product Safety Concerns and Information please contact our
EU representative GPSR@taylorandfrancis.com Taylor & Francis
Verlag GmbH, Kaufingerstraße 24, 80331 München, Germany